The Compleat Chemical Engineer:
A Guide to Critical Thinking

He that hopes to be a good *Angler* must not onely bring an inquiring, searching, observing wit, but he must bring a large measure of hope and patience, and a love and propensity to the Art it self; but having once got and practis'd it, then doubt not but *Angling* will prove to be so pleasant, that it will prove like Vertue, a reward to it self.

Isaac Walton, *The Compleat Angler* (1653)

Robert Barat
Norbert Elliot

New Jersey Institute of Technology

With

Piero Armenante
Anne Buck
Chris Cowansage
Eric Katz
Gordon Lewandowski
Daniel Watts

KENDALL/HUNT PUBLISHING COMPANY
2460 Kerper Boulevard P.O. Box 539 Dubuque, Iowa 52004-0539

Cover photo: View of Laboratory, University of Georgia Biosciences Complex. Courtesy of CRSS Architects, Tom Linehan and Paul Lyle, Visualization Department, Houston, Texas.

This edition has been printed directly from camera-ready copy.

•   •   •

To our students in the Department of Chemical Engineering, Chemistry,
and Environmental Science at the New Jersey Institute of Technology

&

For Katherine Barat and Lorna Jean Elliot

Table of Contents

Detailed Table of Contents

Part 2: The Chemical Engineer in Society

Chapter 8. Engineers and the Environment: Preventing Pollution at the Source by Daniel Watts

Part 3: The Chemical Engineer as Professional Writer

Chapter 9. Communicating Information in Chemical Engineering

Preface

If you enjoy experimenting with new ideas, this book is for you.

The Compleat Chemical Engineer will help you think critically about the situations you will encounter in college and after graduation. This book will also help you communicate the results of your critical thinking—your ideas—to a wide range of audiences.

Your guiding spirit in this textbook is Isaac Walton, a seventeenth-century ironmonger who, although lacking formal education, transformed himself into a man of learning and wit. Friend and biographer of the poet John Donne, Walton is best remembered for his treatise on fishing, *The Compleat Angler*. First published in 1653, the book went through five editions in the poet's lifetime. Its popularity has never waned: it was reprinted 10 times in the eighteenth century, 117 times in the nineteenth century, and at least 40 times in the twentieth century.

Were all these readers interested in fishing? Probably not. Most look to Walton because, in talking about fishing, he is also talking about the human condition. While there are technical discussions—catching the trout and the salmon and the pike—there is also skillful angling for those qualities that make us what we are when we are at our best. With Walton as a guide, we hope to help you find that stream of ideas that will allow you to excel as a professional. Whether you use this book in a single course, in a sequence of courses, or for your own development, we hope that this book becomes a companion that will empower you to be a leader in chemical engineering.

• • •

Organizationally, there are three major parts to this textbook.

In Part I, the topics discussed deal primarily with chemical engineering issues as they are treated in the academy. Chapter 1 will introduce you to our critical thinking model and its applications. Chapter 2 shows you how to approach both industrial and academic chemical engineering problems within an historical context. In Chapter 3, you will see how your senior laboratory course is truly the culmination of chemical engineering, where successful completion of the experiments involves a detailed research plan. In learning how to organize your experimental data and argue in support of your results in Chapter 4, you will see that there are limits to how well something can be known. Finally, Chapter 5 offers important guidelines on how to exploit the rich literature of chemical engineering and related fields.

In Part II, we expand our discussions to societal topics which, while not of an entirely technical orientation, are nonetheless indispensable for the development of a chemical engineer. The often tough ethical issues that chemical engineers face in the work place and in the student laboratory are addressed in Chapter 6. Your perspective toward chemical engineering will be broadened with an architectural view in Chapter 7. Chapter 8 introduces the concepts of pollution prevention, all of which are applicable in the laboratory, plant design, and plant operation. These concepts address the environmental concerns of chemical engineers.

In Part III, we ask you to train yourself as a professional writer. Here you will see that effective oral and written communication is imperative for the successful practice of chemical engineering as a student and as a professional. Chapter 9 shows how research in engineering is linked to technical communications. In Chapter 10, you will learn how to communicate effectively in the academy. In Chapter 11, you will learn how to communicate effectively through various reporting structures commonly used in the organization.

At the end of each chapter, you will find assignments designed to help you operationalize the ideas covered in each chapter. These assignments will encourage you to analyze your own development as a chemical engineering professional.

•   •   •

This textbook was developed in a collaborative program sponsored by the Department of Chemical Engineering, Chemistry, and Environmental Science and the Department of Social Science and Policy Studies at the New Jersey Institute of Technology. Without the vision of Gordon Lewandowski and John Opie, Chairs of these Departments, this work would never have been possible. We also owe special thanks to our Provost, Gary Thomas, for his consistent encouragement of our ideas.

Our debts are great to colleagues who have supported our research and given freely of their time and intelligence: Dana Knox, Robert Lynch, Reginald Tomkins, and Henry Shaw. Paul Zelhart introduced us to the idea of cognitive complexity, and his influence is evident throughout this book. Thanks also to Lorna Jean Elliot for her editorial assistance.

Greater still is our debt to our contributors: Piero Armenante, Anne Buck, Chris Cowansage, Eric Katz, Gordon Lewandowski, and Daniel Watts. Their ideas have enriched the book immeasurably.

And there is the greatest debt of all: the one we owe our students in the Department of Chemical Engineering, Chemistry, and Environmental Science at New Jersey Institute of Technology. Our students were ever patient and gracious in helping us during the six years it took to refine our ideas and to write this book. This work is for them and for those who will follow them in chemical engineering programs across the nation.

•   •   •

Earlier versions of parts of this book were presented and published in the following:

"Programmatic Development of Critical Thinking in the Chemical Engineering Curriculum." *Proceedings, Critical Thinking: Focus on Science and Technology*. Eds. Wendy Oxman-Michelli and Mark Weinstein. 1992. Volume 1. Institute for Critical Thinking: Montclair State College, pp. 496–510.

"Technical Writing in a Technological University: Attitudes of Department Chairs." *Journal of Technical Writing and Communications,* 24 (1991), pp. 411–424.

"Critical Thinking in the Senior Laboratory." *Chapter One: The AIChE Magazine for Students,* (May 1991), pp. 50–53.

"Designing A Critical Thinking Model for a Comprehensive Technological University." *Inquiry: Critical Thinking Across the Disciplines*, 7 (1991), pp. 8–10.

"Implementing a Critical Thinking Model in a Technological University: The Capstone Laboratory." *Inquiry: Critical Thinking Across the Disciplines*, 7 (1991), pp. 11–13.

"Toward a New Paradigm in Undergraduate Chemical Engineering Education." *Engineering Foundation: New Approaches to Undergraduate Engineering Education III*. Banff, Canada. 29 July 1991.

"A Cognitive Model for the Capstone Course: Design, Implementation, and Evaluation." *American Institute of Chemical Engineers*. Chicago, IL., 16 November 1990.

"Hermeneutics and the Teaching of Technical Writing." *The Technical Writing Teacher*, 17 (1990), pp. 150–164.

Academy, Economy & Society: Extending and Supporting General Education. Trenton: New Jersey Department of Higher Education, 1990.

"Integrating Ethics in the Engineering Curriculum: An Innovative Collaboration Between Engineering, STS, and the Humanities at the New Jersey Institute of Technology." *National Association for Science, Technology, and Society*. Washington, DC. 4 February 1990.

Continuing work on pollution prevention is being conducted through funding from Environmental Protection Agency Grant #992903 for Science, Technology, and Society Curriculum Transformation.

1 Thinking Critically in Chemical Engineering

" I cannot, for want of sufficient premises, advise you *what* to determine, but if you please, I will tell you *how*."

Benjamin Franklin in a letter to Joseph Priestly (1772)

Preview

In Chapter 1, you'll have a chance to think about

- the role of the chemical engineer and the necessity for critical thought

- a model of critical thinking for chemical engineers

- the application of that model in various chemical engineering settings

"O Sir, doubt not but that Angling is an Art, and an Art worth your learning: the Question is rather whether you are capable of learning it?

Isaac Walton, *The Compleat Angler*

Part I: A Model for Critical Thinking in Chemical Engineering

Introduction

As you study chemical engineering, you are simultaneously reviewing all that you have learned over the past few years and looking ahead to the future. While there is security in the past, there is uncertainty as well as promise in the future. You will be faced with decisions which must be addressed on many levels: technical, ethical, cultural, personal.

As a chemical engineer, you will face the growing demands of international competition as the United States becomes increasingly integrated into a global market place. As a nation, the United States will face unprecedented challenges from a United Europe and the Pacific Rim countries; uncharted opportunities await in the new nations of the former Soviet bloc.

Bound together with these economic challenges is the rapid rate of scientific discovery and technological development. While the basic principles of science and engineering do not change, their application to the generation of new knowledge and technologies is almost boundless.

The Committee on the Education and Utilization of the Engineer reminds us that international economic pressures and the demands of the technology explosion must be recognized.[1] We believe that you can prepare yourself for this world by becoming a chemical engineer dedicated to critical thinking. In your courses, we encourage you to take an active role in the development of the critical thinking skills which you will need for future success.

To build these skills, you will need a model to follow.[2] This model must be valid, rich, and appropriate for the discipline of chemical engineering. With an engaging model, critical thinking will become a valuable part of your education as a chemical engineer.

A Critical Thinking Model

The model we will use in this textbook is shown in Figure 1.1 below.

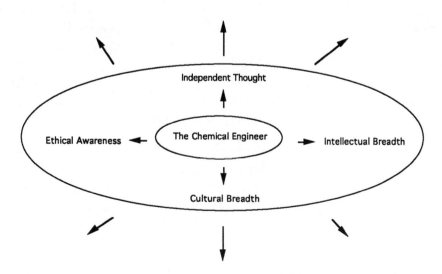

Figure 1.1 A Critical Thinking Model for Chemical Engineering Students

At the core of the model is the **Chemical Engineer**. Such an individual, we believe, incorporates the ideal of the universal thinker within the context of our discipline.[3] This individual is neither a technical specialist working within a cultural and ethical vacuum nor a social revolutionary who lacks technical competence.

A chemical engineer who is dedicated to thinking critically works toward the four broad goals of our model:

Independent Thought—the ability to move beyond simply memorizing formulae to analyzing and constructing meaning from complex data and theory;

Intellectual Breadth—the ability to move beyond disciplinary specialization to look at issues from different perspectives;

Cultural Breadth—the ability to understand differences in perspectives arising from economic status, race, gender, and class; and

Ethical Awareness—the ability to understand the impact of our technical decisions on ourselves, others, and the environment.

The means by which you strive toward the goals of critical thinking in chemical engineering can be developed for specific tasks and courses. For example, the critical thinking skills needed for successful completion of a design course are somewhat different from those for the capstone laboratory course. However, on a fundamental level, the critical thinking skills for each are similar and complementary. As you will see, these skills are all consistent with the broader goals of the base model shown in Figure 1.1.

We will now analyze this model more closely. We will illustrate its adaptability through examples drawn from industry and the senior chemical engineering laboratory course.

The Core: The Compleat Chemical Engineer

The image of a Renaissance intellectual evokes for us individuals of diverse talents in science, literature, and the arts. The European Renaissance—a period we may date from approximately 1350 to 1600—is generally considered a time of rebirth. Those living during that age believed that they had restored learning that had been dormant during the eleventh, twelfth, and thirteenth centuries. The Renaissance witnessed the invention of printing, the resurgence of economic prosperity, and the presence of Leonardo (1452–1519), Erasmus (1466–1536), Michelangelo (1475–1564), and Shakespeare (1564–1616). In their humanism, Renaissance thinkers advocated living virtuously; indeed, the period has been referred to as one of civic humanism. Individuals became involved in applying their education for the good of society.[4]

The Chemical Engineer at the core of our model embodies a kind of Renaissance spirit. The Chemical Engineer is an intellectual who has a broad, multidisciplinary education. Of course, the engineer has a firm, working grasp of the fundamentals of chemical engineering and, perhaps, specialization, such as reaction kinetics or transport processes. In addition, the engineer has a reasonable grasp of other engineering fields and the sciences such as physics, chemistry, and mathematics. Here is an illustration of how this unique student will function in the laboratory:

Student Group #1 is preparing to operate the gas absorption experiment. An analyzer is available to measure the concentration of ammonia in the gas streams. The analyzer's appearance is literally that of a black box.[5] Data collection begins, and the group leader notices confusing, fluctuating readings. One of his partners observes an alternately flashing light corresponding to the fluctuating readings. She suggests that the operating manual for the analyzer be obtained and that the front panel of the analyzer be opened. In this way, she argues, the group could determine how the instrument really works and, perhaps, explain the fluctuating readings. The third group member strongly dissents, saying that such action is beyond the scope of this lab course. "This is not an instrumental course. Let's just ask the teaching assistant; maybe he knows." The group leader, however, is not afraid of extending beyond the confines of chemical engineering, but is nevertheless aware of the need for safety. He consults with the professor and obtains approval for examination of the inside of the "black box." After a short period, the group discovers that the analyzer operates on the principle of infrared absorption, which they recall from physical chemistry class. They also realize that the fluctuating readings are the result of a mechanical selenoid switching valve. The proper settings are now made, and the analyzer generates reliable data.

As the example illustrates, the promising chemical engineering student does not function in the vacuum of equations and calculations.

But what about the role of critical thinking for chemical engineers *after* graduation? Following the model, these chemical engineers become humanists concerned with the impact of their decisions on others and the environment. These chemical engineers examine the past in order to avoid the repetition of previous mistakes. Thoughtful researchers who recognize that chemical engineering must extend beyond the laboratory, the office, and the plant into the community, chemical engineers also know that oral and written communication skills are essential. Here is an example:

The environmental engineering section of the large chemical plant has been directed to design an on-site hazardous waste incinerator. Various operating modes are to be considered: start-up, full capacity, upsets. Assignments are given, and the designs completed. The engineers meet to discuss how to present the results. When the issue arises as to how to "sell" the incinerator to the local surrounding community, one engineer responds by saying "That's not our problem; that's a job for the lawyers and the PR guys. We only care about the nuts-and-bolts design." A colleague disagrees. In addition to his design results, he has compiled statistics on emissions from other industries in the areas. He has also performed a risk analysis of the various incinerator operating modes. His conclusion is that the incinerator will be a good neighbor for the community, generating electricity and steam, with the odds of a major emissions incident being quite small due to design safeguards. Most importantly, he has translated the results of his technical research for presentation at a local town meeting. The facts are given in a clear, honest manner using language which lay persons will understand; of course, the presentation is informative, not condescending. Months after the presentation, the town council approves the incinerator without significant popular opposition.

Thus, you can see that the skills of the informed chemical engineer are exhibited on both student and professional levels. Let us now examine the broad goals of the critical thinking model.

The Goal of Independent Thought

You must be able to think critically in order to show independent thought. For example, when presented with complex laboratory data from which you must determine important results, you should avoid rifling through textbook after textbook looking for that one "magic" formula that will reduce the data in one step to the desired results. Following the spirit of the critical thinking model, you should instead critically examine the data in conjunction with the experimental apparatus. You should determine the important physical phenomenon and mechanisms which apply during the experiment, then write or determine the governing equations and relationships. This process might involve examination of textbooks and literature references to find an analog based on your education and experience. You will be motivated to construct meaning, not simply to "plug" data into a formula.[6] Here is an example of what we mean:

> A student lab group performs an experiment with a non-adiabatic, non-isothermal batch reactor. First, a cooling curve is generated from simple hot water. Next, an exothermic hydrolysis reaction is performed, and a temperature-time curve is obtained. When faced with the task of obtaining chemical reaction kinetic parameters, the group wonders what to do with their data. One student says: "This is temperature data. We can't get kinetics from this. I'll go check the literature reference." Upon doing so, the group finds other research that appears close to what they have found; the literature also has provided accompanying equations. The group simply copies these equations and blindly plugs in the laboratory data. The group subsequently generates kinetic parameters which vary significantly from accepted literature values. In the report, the group blames faulty equipment and "experimental error." The professor examines their theory and determines that the group had used equations for an adiabatic system. The group had not realized that their system was not insulated since they had given no thought to the physical processes occurring and how these processes should be accounted for in the analysis of the data.

In order to exercise independent thought effectively, you must have the kind critical daring that the above group lacked.

The Goal of Intellectual Breadth

The capacity for intellectual breadth means that you must go beyond the confines of your chemical engineering specialization. When you have completed all the prerequisite courses for the undergraduate chemical engineering degree, you will have at your disposal all the fundamentals of chemical engineering: fluid mechanics, heat and mass transfer, reaction theory, and thermodynamics. However, you should be confident enough to generate an integrated understanding through multiple perspectives derived from other disciplines. Here is an example from graduate school:

> A chemical engineering graduate student wondered how he would approach his assignment. He had to determine how closely a backmixed reactor approached perfectly stirred behavior. He had to develop a system to diagnose the reactor based on laser light scattering. "I'm a chemical engineer," he first thought. "This is a job for a physicist." With time, though, the student developed the diagnostic laser system. When a severe spurious light background problem threatened the entire project, the student initiated an interdisciplinary approach: "I can tackle this problem. All I need to do is to optically subtract the background." He determined the necessary criterion for this operation; he then sought out electrical engineering graduate students to assist him in the circuit design. He independently developed a calibration procedure for this system. After all this non-chemical engineering preparatory work, the actual reactor experiments began. Fascinating, unexpected data were obtained which became the basis for several journal articles.

Intellectual breadth means that you are not afraid to go outside of your discipline. You must have sufficient knowledge to define the problem and what it requires, but you must also be willing to consult those in other disciplines and fields. If they are true professionals, these colleagues will share your enthusiasm for tackling a technical problem through multiple perspectives. Technical problem solving, particularly in industry, involves multidisciplinary teams.

The remaining two goals involve capacities which are generally not technical in themselves. However, as we explain below, it is clear that chemical engineers cannot operate in an ethical or cultural vacuum.

The Goal of Cultural Breadth

As the world becomes more and more a global community, cultural tolerance is increasingly critical to the well-being of that community. Chemical engineers have a two-fold responsibility as contributing community members: first, to embrace all that various cultures can offer to strengthen the community; second, to apply this knowledge toward the safety and well-being of all members of the global community.

As an increasing number of students from other cultures come to learn in our culture, the exposure of chemical engineering students to other global community members is inevitable. This is positive insofar as it enables you to begin thinking in terms of the common efforts and understandings of the community at large. It is also positive insofar as such ethnic diversity realistically reflects the demographics of the working industrial community.

As chemical engineering students, then, it is important to cultivate an international perspective that is not driven strictly by economic or nationalistic considerations. Below is an example of an industrial dilemma that requires global community considerations for its solution.

> A chemical company in New Jersey is faced with a dilemma. New state and federal regulations taking effect in one year will make it illegal to landfill hazardous wastes generated by its plant. Plant managers and senior engineers gather to discuss various options. The least costly option, at least in the short term, is to ship the waste to a small third-world country with a history of quietly accepting outside hazardous wastes at relatively low costs. After extensive research, one engineer declares his opposition, saying that the landfills there are poorly managed, leak badly, and threaten local farmers and their families. He recommends that a long-term perspective should be adopted. Plant modifications should be made to reduce waste generation. The remaining waste should be treated in on-site units, producing acceptable, even saleable, by-products. While this option is more costly in the short run, respect for another culture is upheld and the problem is solved for the long term.

To understand cultural diversity, often you will have to analyze the historic implications of a problem by means of the analytic methods discussed in Chapter 2. The engineer in the example above brought an historical perspective to the problem. The above example also has implications for our final goal, the capacity for ethical awareness.

The Goal of Ethical Awareness

Serious consideration of engineering ethics is long overdue. Whether in the laboratory, the design room, or the plant, the chemical engineer is often faced with ethical challenges. For example, although it may surprise you, the student laboratory is a center for intense ethical decision-making. Your reaction to the example below might offer some insight as to your own sense of engineering ethics:

A student group completed their experiment on fluid flow and began data analysis. The group plotted the experimental friction factor versus Reynolds Number for flow in a pipe with six points in the original data set. The regressed line showed a slight upward trend, although the literature curve angled downward. One student expressed dismay: "Our data indicates a trend opposite of that found in the literature. We're in trouble." Another suggested a fix: "Let's drop the two outlying points which cause the upward trend. Nobody will know the difference." The group leader replied that there was no justification to throw data away. She recommended that the group re-examine the data collection technique. "Maybe we can explain these points," she argued. "The professor won't penalize us for discussing our results critically."

The ethically-conscious chemical engineer will not make expedient engineering decisions merely to obtain desirable results, a topic of Chapter 6. You must instead use all your capabilities to understand the unexplained result, the "outlying data point," that might, upon critical examination, reveal promising and satisfying information.

Now that the broad goals of our critical thinking model have been discussed, we will examine the ways you can achieve these goals within the following settings:

- introductory courses in chemical engineering

- courses in plant design, and

- laboratory courses.

Part II: Application of the Model for Beginning Chemical Engineers

Pursuing the Goals of Critical Thinking Within the Material and Energy Balance Course

We now turn to a brief discussion of the critical thinking skills identified with successful performance in chemical engineering material and energy balance courses. In the academy, these courses are most often given in the sophomore year, and thus they often serve as an introduction to the field.

The critical thinking skills needed for such introductory work are shown in Figure 1.2 as a different outer shell around the same core and broad goals of our critical thinking model.

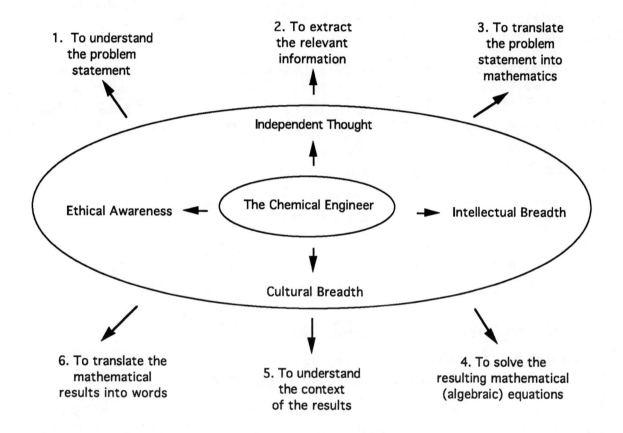

Figure 1.2 A Critical Thinking Skills Model for Chemical Engineering Material and Energy Balance Classes

As you will see later in this chapter, these six skills—critical to your success in your introduction to the discipline—are consistent with those skills identified as critical to the more advanced plant design course and to laboratory work. You need to develop the core elements of critical thinking early in your course work and apply them systematically throughout your university education and your career. In the following sections we will discuss each of these important skills independently.

Understanding the Problem Statement

Real problems are always presented in words, either written or spoken. In either case, the chemical engineer must comprehend the problem statement, rephrase it, and return the revised problem statement back to the source for confirmation. In many cases, the original problem statement might be incomplete (underspecified), or have too many constraints (overspecified). Either of these cases will require clarification before a problem solution can be attempted.

For example, imagine that your supervisor directs you to buy a new car. He asks you how much it will cost and thus presents a problem statement. This problem is nevertheless underspecified since a complete description of the kind of car required—compact or luxury, automatic or standard, passenger side air bags or not—is missing.

Conversely, suppose your supervisor directs you to buy a new car for $5,000. The car must have eight cylinders, automatic transmission, air conditioning, and a compact disc player. This problem is overspecified since too many constraints are stated. One cannot purchase a new car today with all these features for only $5,000. Therefore, you must request that your supervisor prioritize the constraints.

These two examples also have analogies in algebra, a form of critical thinking central to success in material and energy balances. To obtain a unique solution, the number of unknowns must equal the number of available equations. The problem can be neither underspecified (an infinite number of possible solutions) nor overspecified (no possible solutions).

Extracting the Relevant Information

In any real problem, there is always relevant and irrelevant information presented. You have to determine which information is relevant in order to formulate a solution.

For example, your supervisor asks you to buy a new car on Tuesday for $12,000. It must have six cylinders, an automatic transmission, air conditioning, and a red exterior. Are "Tuesday" and "red" significant? If not, you will have to rephrase the problem to include only relevant information.

Translating the Problem Statement into Mathematics

This is *the* essential engineering exercise. The problem statement in words must be translated into mathematics; the statement and its translation can either be trivial or complex.

For instance, your supervisor asks you to determine the size of equipment needed to manufacture 1,000 pounds per hour (pph) of nylon. Last year, you specified such equipment for making 500 pph of nylon. Therefore, the answer to the new question might be reached simply by multiplying the old size by a factor of two. However, if she asks you to size equipment for making 5,000 pph of polystyrene, the extrapolation from 500 pph of nylon would be much more complex.

So, beginning students must remember that the number of available independent material balance equations is equal to the number of components for each process step. There is also an independent energy balance for each process step. The number of unknowns must be obtained from a translation of the problem in words. You have to be sure that the number of available equations equals the number of unknowns; otherwise, there is no point attempting a solution.

Solving the Resulting Mathematical (Algebraic) Equations

Engineering is the organization and extrapolation of experience in a mathematical context. By the sophomore year of a chemical engineering program, you will be expected to solve linear algebraic equations. If you have more than three simultaneous equations to solve, a numerical technique is required.

Understanding the Context of the Results

Chemical engineers are not asked merely to solve algebraic equations. Instead, they are asked to solve problems in a broader sense. What is the significance of your mathematical results to your employer? In order to answer, you need to understand the *context* of the problem. That understanding distinguishes a professional (compleat) engineer from a technician.

Translating the Mathematical Results into Words

In order to succeed in your organization, you must be able to explain to people, especially your superiors, the significance of your mathematical results. What do your results really mean in physical and economic terms? The ability to translate your results into words is closely tied in with recognition of the context of your work. Part 3 of this textbook—Chapters 9, 10, and 11—is dedicated to communication.

Part III: Application of the Model in Plant Design

Pursuing the Goals of Critical Thinking in the Design Class

We now briefly consider a similar analysis of the critical thinking skills needed for chemical engineering design work. These skills are shown in Figure 1.3 as a different outer shell around the same core and broad goals of our critical thinking model.

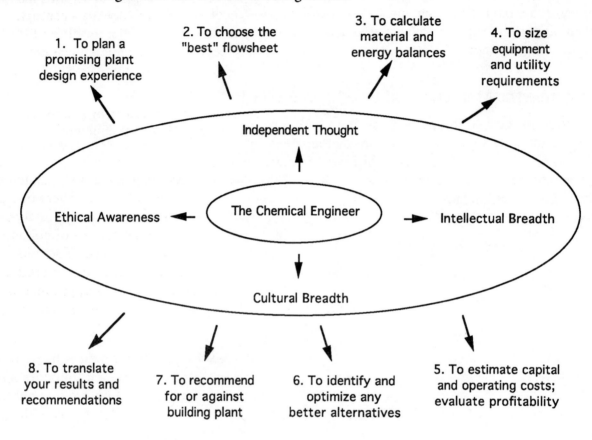

Figure 1.3 A Critical Thinking Skills Model for the Chemical Engineering Plant Design Class

It is important for you to take a moment to consider these eight critical course objectives. It is also important to note their relationship to the six objectives of the Material and Energy Balance course previously discussed and outlined in Figure 1.2 before we begin a discussion of each point independently.

Planning a Promising Design Experience

Effective planning is required for plant design, one of the most complex processes in chemical engineering. Design, the planning process of engineering, is often most misunderstood by its very practitioners. Taken as a static concept, design is often thought to be only an analytic step. This position is sharply criticized by the late T. K. Sherwood of the Massachusetts Institute of Technology:

> The typical engineering accomplishment of importance results from a sequence of (a) recognition of a social need or economic opportunity, (b) a conception of a plan as to how the need may be met or the opportunity seized, (c) an analysis of the merits of the conception and the consequences of proceeding with the plan, (d) the building or construction of the machine, plant, or bridge as conceived, and (e) its operation. Item (c) is a sort of go no-go step; if the analysis is discouraging, the conception step and the analysis are repeated.

> Engineering science is a phrase which has come into use in recent years to describe the *analysis* step. Good engineers today must be exceedingly proficient in analyzing a concept; quantitative mathematical analysis based on first principles is one of the good engineer's most powerful tools. But it no more constitutes *engineering* than any one of the other four activities listed. In particular, it must be noted that conception ("design") comes before analysis; the analyst must have something to analyze.

> As technology develops, it seems likely that those engineers who can recognize social needs or economic opportunities and who can conceive and plan will be [those] most sought after. [7]

First, Sherwood recognizes design as a multifaceted process. He recognizes that technology must be defined through its technical, organizational, and social aspects. Second, Sherwood poses an implicit warning: if design becomes nothing more than a kind of decontextualized analysis, the essence of engineering—application in context—will be lost. The basic guidelines of collaboration therefore hold true, as well as the identification of the tasks to be performed. You must establish the goals of the design—What do you wish to produce and how much of the product is necessary? You must establish the limiting criteria such as feedstock availability, product marketability, environmental constraints, and societal concerns. Only after your objectives and limitations are critically established can you proceed.

The literature search is also important here. A component of the search which is likely to be absent from the literature search before an experiment is the search of the patent literature. Often individuals or organizations will file for and/or obtain a patent for a process before it is published in the general scientific or engineering literature. Chapter 5 is devoted to helping you with such searches.

Choosing the "Best" Flowsheet

After planning the design and searching the literature (including a patent search), several possible flowsheets should be assembled. Each should be broadly critiqued based on such issues as availability of technology, licensing, environmental and cultural impact, availability of feedstock and utilities, availability of markets for the products, and so forth.

Some alternatives might be rejected outright for an obvious reason, such as the lack of a product market. To critically evaluate the others, you should employ a rating system in which each major consideration is given a weighting. The flowsheet with the best overall score will be used in the plant design which follows. Keep the other alternatives and their rankings, though. If the top candidate is rejected later on in the design process, you must go back and consider the other flowsheets. This process of selecting criteria and ranking flowsheets exemplifies a key aspect of critical thinking in plant design.

Calculating the Material and Energy Balances

No plant design can be legitimately performed without material and energy balances. The overall process is likely to consist of a complex set of unit operations and reactors. A generic flowsheet might include feed preparation, reactor, separation, and recycle.

The calculation of material and energy balances is greatly facilitated by any one of a number of computer process flowsheet simulators which are quite common and readily available. These are often quite user-friendly, providing you with considerable flexibility. However, even the best of computer simulators will not yield good quality results if the user-supplied input information is poor. Users should strive to estimate as much key stream information as possible. For a multi-component system, simplifications can often be made which will provide a decent "first cut" as computer program input. Critical thinking is important to making good estimates. In fact, since so many of the difficult material and energy balance calculations are eliminated by computer simulators, generating good estimates based upon critically valid assumptions is actually the most difficult technical task in plant design.

Sizing Equipment and Utility Requirements

The amount of required utilities such as cooling water, steam, and electricity, and the sizing of equipment such as reactor vessels and distillation columns often involve additional estimating which is enhanced by critical thinking. The material and energy balances will provide flowrates and heating/cooling loads. However, typical design practice at such an early stage usually calls for a degree of oversizing to compensate for unforeseen contingencies. This generally provides for a more conservative design and cost estimates. Critical thinking is required here. Oversizing should be prudent but not wasteful. It should be applied where your knowledge of the material and energy balances is weakest (for example, in side reactions, separation efficiencies, and so forth). Both optimistic and wasteful oversizing can lead to choice of the wrong plant design.

Estimating Costs and Determining Profitability

A particular plant design pursued to this stage is ultimately made or broken here. You must decide upon an economic basis (e.g. net rate of return on investment) for the cost estimate, and you must determine both capital and operating costs. If the proposed design paints a financial picture consistent with your objectives, then it can be pursued to a higher, more detailed level. If not, then the design should be rejected.

Like sizing at this early stage of plant design, cost estimates must not be too optimistically low or conservatively high. Either scenario does an injustice to the particular flowsheet and can lead to incorrect conclusions. A significant ethical component becomes active at this stage. Because profitability is a major (though it should not be the only) factor in the ultimate decision regarding a particular plant design, cost estimates could be used to pursue an individual's particular agenda. For example, a middle manager who has made unsubstantiated claims to upper management finds that the cost figures provided from below disprove his claim. Instead of admitting his error, the middle manager changes the economic basis from a reasonable one to one which is highly speculative. He manipulates the capital and operating costs until the "correct" answer is obtained. Such behavior clearly would not be exhibited by the critically thinking chemical engineer.

Identifying and Optimizing Promising Alternatives

If your design has not met the established profitability and other objectives, you must pursue alternatives. Begin by referring back to the earlier ranking of flowsheets to determine the next "best" one, and begin a second round of work. Informed by the knowledge and insight gained from your first round, your second pass through the above steps should be of a higher quality, your cost estimates better informed.

Recommending For or Against a Plant Design

Once a particular plant design has achieved all your objectives, you must make a recommendation to your superiors, usually higher-level managers. In effect, you must now make an argument firmly grounded in critically evaluated material and energy balances and cost estimates. Just as in the laboratory, you must be prepared to answer any challenge to your results. You must be willing to discuss openly where your estimates are not particularly strong and their impacts on the conclusions.

Translating (Communicating) Your Recommendations

Communication is essential to the successful completion of your plant design. You must get the message across. Proper translation of your work and results for the particular audience in question is critically important for successful argumentation. Since upper management cannot afford the time to look over all your calculations and assumptions, you must get to the "heart of the matter" in language and in a context which are appropriate. What a loss it would be if, after all your hard work, your results were rejected simply because you did not adequately present them. Chapter 11 will help you become familiar with the forms of communication commonly used in the modern organization.

Part IV: Application of the Model in the Laboratory

Pursuing the Goals of Critical Thinking Within the Laboratory

In the laboratory, there are means by which you can strive toward the goals of critical thinking. There are seven characteristics or skills which you can develop. They are described in Figure 1.4:

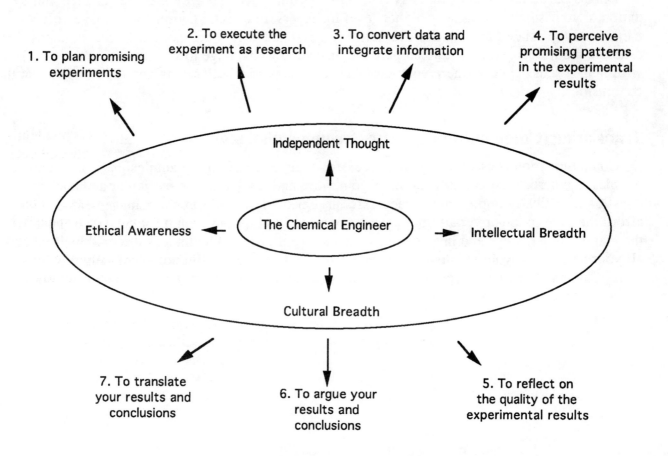

Figure 1.4 A Critical Thinking Skills Model for the Chemical Engineering Laboratory

These are the tools of the critical thinking model within the laboratory. As your career develops, these skills will grow with you. They will become, in effect, the means by which you will approach and tackle all problems.

Take a moment to recognize and think about each of these seven skills. Relate them to the six skills outlined for your Material and Energy Balance course (Figure 1.2) and the eight skills outlined for your Plant Design course (Figure 1.3). Ponder the relatiohnship in all three figures between the core critical thinking model and the specific course objectives. Each objective allows for a manifestation of critical thinking.

These seven skills, helpful in attaining critical thinking ability, will be illustrated primarily by way of a liquid-liquid extraction student experiment described here:

> In this experiment, acetic acid is extracted from n-hexyl alcohol with water. A reciprocating Karr column is used to promote contacting of the phases. The organic feed is fed into the column near the bottom, just below the lowest oscillating plate. Water is fed near the top, just above the highest plate. The raffinate phase is withdrawn at the very top, with the acid-rich extract taken from the very bottom of the column. The feeds are delivered with diaphragm metering pumps. A speed-controlled motor oscillates the plate array. Feed and outlet streams are analyzed for acetic acid by titration with base. The objective of the experiment is to evaluate the extraction efficiency of the column as a function of the plate oscillation rate.

Planning Promising Experiments

The effective planning of a promising experiment requires that the important tasks be identified. You must select the tasks based on criteria of effective use of time, utility of obtained data, and safety. These data must be directly useful in meeting the stated objectives.

An equally important component of planning is collaboration, discussed in Chapter 3. You must work well within the group, completing your assigned tasks. While division of responsibilities is key, effective collaboration extends beyond the actual experiment to the data analysis and report preparation stages. The literature search, discussed in Chapter 5, is also part of the planning stage. Here is an illustration of effective collaborative planning from the liquid-liquid experiment:

> The group appoints you as leader. You identify the important experimental tasks for the group: a) calibrating the feed pumps, b) setting the feed rates, and establishing and maintaining a steady-state column operation, c) measuring the plate array oscillation rate, d) collecting samples, e) analyzing samples by titration, and f) generating a working plot of acid concentration in the extract versus oscillation rate. You estimate the time requirements for these various tasks; then you assign your group members accordingly. Work begins.

Executing the Experiment as Research

It is important that the experiment be executed with the attitude that your group is performing a research activity. (A model for chemical engineering research is provided in Chapter 3.) Do not look upon the experiment as simply an exercise in validating equations. You will probably discover that the actual experiment does not behave according to the neat, simple relationships you learned in prior classes. You should thus approach each experiment as if you are the first ever to perform it. Let's continue with our example:

> After calibrations and the achievement of a steady-state column operation, sample collection begins. Titrations of the first extract samples reveal a significant concentration of acetic acid. The oscillation rate is increased, and new samples are taken. Titrations show an even greater acid concentration in the extract. A working plot is started of acid concentration versus oscillation rate. The group speculates that this trend will most likely continue. Later, the group notices that at a high oscillation rate the liquid in the column no

longer appears as drops of water dispersed in organic, but looks milky. Members wonder whether to continue. After more observation and discussion, they suspect that an emulsion has occurred. The phases are no longer separate, so they conclude that extraction is no longer feasible. The experiment is stopped and the column drained.

Notice that the group did not consult textbooks to decide what to do. The literature might have warned them about the phenomenon, but it was up to them to recognize it. Using observation and discussion—crucial elements of a research orientation—they reasoned that the fundamental mechanisms which had been governing the column no longer applied during the emulsion.

Converting Data and Integrating Information

Students often exhibit the tendency for what we call the "data dump." All data—from raw quantities such as titrant volumes, to final calculated values such as the Height Equivalent of a Theoretical Plate—are lumped together in the report for the reader to decipher. Because discrimination of data depends partly on the theoretical and semi-empirical relationships chosen or derived, assigning degrees of importance to various raw data and calculated quantities will help you organize your thinking. In Chapter 4 we will talk about the necessity of recognizing various data levels before converting raw data into final results.

We have noticed that some students tend to minimize a laboratory experiment simply because of its title. For example:

> In a recent exothermic reaction kinetics experiment, we noted the response of the student group when it was suggested that their temperature/time data might be a convolution of the true reactor temperature history and the time response function of their sheathed thermocouple. "That's a process control topic," one student said. "This isn't that course. Besides, this is a reaction kinetics experiment."

Clearly, the students were unable to integrate information derived from their previous courses. (This is a common problem for incoming chemical engineering seniors).

Perceiving Promising Patterns

Over the years, we have observed that many students will take their experimental data and correlate them using a theoretical or semi-empirical relationship without considering whether that relation is valid for the experimental conditions and apparatus employed. When subsequently faced with a discrepancy between their results and literature predictions, typical student responses are: "We must have done something wrong" or "Experimental error is the problem."

You should be prepared to bring all your engineering and scientific training to bear on your experimental data. Do not just run to your course textbooks and search wildly for the "right" equation. Here is a good example of thoughtful recognition of experimental patterns:

> A laboratory group examines its extraction data and ponders what model to use. One student says: "Let's first look at the equilibrium data in the literature. This will tell us if a dilute system assumption is valid." Another recalls that "this column is a stage process, so we can derive a simple operating line." "Great!" the leader says. "Now we can step off theoretical stages. HETP from that point is a snap. " The first student then adds: "You know, the agitation makes the bubbles smaller, so I'll bet the contacting is better with faster agitation. We already know that the acetic acid concentration in the extract goes up at higher agitation rates. That means the number of stages should increase."

This group has developed an entire approach to modeling data by considering fundamentals from their previous mass transfer class and from their present laboratory observations. Note that the group's perception of a recognizable pattern was completed before a mass transfer textbook was consulted.[8]

Reflecting on the Quality of Your Results

Here is a recurring question that instructors direct at students presenting their laboratory work: "How good are your results?" Often the students do not have an answer. If their model prediction does not align itself precisely through the experimental data points, students usually assume a penitential posture or blame the equipment. The quality of your responses to questions of experimental confidence is an important issue. Why? Experimental confidence has a direct bearing on how you draw conclusions.

As students, you generally do not have sufficient time in a chemical engineering laboratory course to obtain statistically large amounts of data. For example, in titrating samples from the extraction experiment discussed earlier, if time allowed, the students would take a large number of samples (e.g. 10) for each run and titrate each. Then, a mean concentration and a standard deviation could be determined for each run.

Nevertheless, estimates can be made as to the precision of various measurements. A simple additive uncertainty approach using partial derivatives can be used, for example, as described in Chapter 4. In this way, uncertainty ("error") bars can be calculated for the final results.

Let us rejoin the group who performed the experiment in fluid flow:

> Having satisfied themselves that their data collection techniques were acceptable, the members of the group took a second look at their plot of Friction Factor versus Reynolds Number for the straight run pipe. "We're not throwing away any points," the leader said, "because, along with the ethical problems that will arise, there aren't that many data points to begin with. Let's calculate error bars." The leader then assigned new tasks. One student returned to the lab and estimated the precision of all the measurement devices used. Another derived expressions for error bars in the Reynolds Number and Friction Factor using partial derivatives. The third combined everything and generated error bars in both the x and y directions for each point. The group was very surprised to see that the uncertainty region around each point was quite large. They realized now that a line could be drawn through these regions that would trend in the same manner as the literature curve. They decided to present the plot with the points, an "eyeball best fit," the uncertainty bars, and the literature curve.

Note that the group's confidence was increased through additional analysis, not through blame or haste.

Arguing Your Results

Students and professionals should be prepared to argue effectively in support of their results and conclusions. Effective and convincing argumentation involves more than good technical writing or oral presentation techniques. It requires a firm grasp of what has been accomplished, as well as a critical attitude toward results.

You can argue strongly for your conclusions if they are based on sound technical facts, data, and inferences. If not, then your argument is a mere "house of cards." As a critically thinking student or professional, you must also remain receptive to the comments and suggestions of others, even if they are contrary to your position. A mind that is thinking critically, as the example below suggests, is always an open mind:

> A student group had evaluated the performance of a gas absorption packed column. From their data, they had determined overall and film mass transfer coefficients for the absorption of ammonia from air into water as a function of gas velocity. Their plot of gas film coefficient versus gas velocity was presented in their report with uncertainty bars around each point and a similar curve of literature values. Their "best fit" curve through their points did not correspond to the literature curve. They correctly pointed out that both curves followed the same trend and that the literature curve fell within the uncertainty bars around more than half of their points. From a position of quantitative strength, they argued that, within experimental uncertainty, their results corresponded well with accepted literature correlations.

While we will have more to say about argumentation in Chapter 4, notice that the group offered a qualified claim supported by data. By means of a comprehensive argument, their results were convincing.

Translating (Communicating) Your Results

To achieve the goals of critical thinking, you finally must be able to communicate your work. The ability to translate results for any audience can actually be defined as the key to effective technical communication. In order for your message to be conveyed, you must tailor the presentation, whether oral or written, to the audience. Failing to communicate your results is equivalent to not having done the work at all. Communication exhibits your critical thinking ability.

The importance of effective communication of your work cannot be overemphasized. Your group may have spent many hours of hard work evaluating your data and generating results. You might even have discovered something new and unexpected. However, without effective communication, your work will have been for naught because the audience, whether listening or reading, will not appreciate it. Translation (communication) is so important that a major portion of this book—Part 3—is devoted to it. If you practice techniques of effective communication, you will be able to avoid the situation described below:

> In an oral presentation to an audience of interdisciplinary faculty, a student group presented their results and conclusions regarding the effects of agitation on extraction efficiency in the Karr column. The group showed beautiful plots of HETP versus agitation rate. One faculty member from the Department of Mechanical Engineering asked this question: "Your results are impressive, and you appear to have worked quite hard. But I'm still having a tough time understanding why agitation helps extraction column efficiency. Can you help me understand this process?" One student answered that "the column is more efficient at higher agitation because the number of theoretical stages increases." The mechanical engineering professor replied: "Fine, but I don't really know what you mean by a stage. Can you give me a physical picture of the function of the agitation?" The group was dumbfounded.

The student group above could not provide a physical picture, unencumbered by equations, of what occurred in their experiment. As a judge of the final oral presentations in the capstone unit operations laboratory, the mechanical engineering professor could not award the group a top score because the students were unable to translate their work. He felt that they really didn't understand or appreciate what they had done. Providing a simple physical picture or an analogy, as we discuss in Chapter 11, is one of the most effective translation devices to ensure effective communication and to demonstrate your own understanding of what you have accomplished.

Closing Comments on Chapter 1

We opened this chapter with a discussion of the design of a critical thinking model. We then illustrated how this model can be implemented in chemical engineering classes and laboratories. This process shows the interrelationship of theory and practice. Without a comprehensive and appropriate theory, classroom activities become mere skills without creativity.[9] Without an articulated classroom model, the theory becomes vague and meaningless.

We want to close this chapter with a challenge: it is up to you to implement the critical thinking model. It is necessary that you take charge of your own intellectual development. Those of our students who have achieved ownership of this model have been rewarded by it. They spend countless hours revising reports and preparing for oral presentations—they become collaborators in their own intellectual development—precisely because they realize that the model will transform them into aggressive, competent, promising professionals.

Notes and References

1. Committee on the Education and Utilization of the Engineer, Commission on Engineering and Technical Systems, National Research Council, *Engineering Education and Practice in the United States: Foundations of Our Techno-Economic Future*, Washington, DC: National Academy Press, 1985.

2. For a complete description of the model and its sources, see the following article: Norbert Elliot and Robert Barat, "Critical Thinking in the Senior Laboratory," *Chapter One: The AIChE Magazine for Students*, May 1991, pp. 50–53.

3. Following Stephen Toulmin in *Human Understanding* (Princeton: Princeton University Press, 1972), we use the term discipline throughout this textbook to refer to the activity of chemical engineering. The components of a discipline, as Toulmin points out, are numerous: "a body of concepts, methods, and fundamental aims"; "a communal tradition of procedures and techniques for dealing with theoretical or practical problems." As well, Toulmin includes "the current explanatory goals," "the current repository of concepts and explanatory procedures," and the "accumulated experience" of those working in the discipline. For a further discussion of the meaning of the term "discipline," see Lindley Darden and Nancy Maull, "Interfield Theories," *Philosophy of Science*, 44 (1977), pp. 43–64.

4. For an excellent discussion of the Renaissance that has informed our own, see the following textbook: Jackson J . Speilvogel, *Western Civilization*, I, New York: West Publishing Company, 1991. Chapter 13, "Recovery and Rebirth: The Age of the Renaissance," pp. 402–442.

5. "The word black box," writes Bruno Latour in *Science in Action* (Cambridge: Harvard University Press, 1987), "is used by cyberneticians whenever a piece of machinery or a set of commands is too complex. In its place they draw a little box about which they need to know nothing but its input and output" (pp. 2–3). Nevertheless, "Some little thing is always missing to close the black box once and for all" (p. 13). In many ways, our critical thinking model is designed so that you will, metaphorically, be willing to open a few more black boxes.

6. For a fascinating account of how meaning is socially constructed among groups by means of writing, see the following: Bruno Latour, *Laboratory Life: The Construction of Scientific Facts*, Princeton: Princeton University Press, 1986.

7. T. K. Sherwood, "The Teaching of Process Design," 1959. Rpt. in *Chemical Engineering Education,* Summer 1985, pp. 121–123, 164. For more on design, see also the following: D. N. Perkins, *Knowledge as Design*, Hillsdale, NJ: Lawrence Earlbaum, 1986.

8. The ability to recognize patterns of thought, a concept taken from Gestalt (meaning "configuration") psychology, cannot be overstressed in its importance to critical thinking. Our devotion to this concept has resulted in the examples that illustrate this textbook. By recognizing holistic patterns of effective performance in chemical engineering, you will be able to incorporate them in your own work. For a classic yet popular introduction to Gestalt psychology, see the following: Wolfgang Köhler, *Gestalt Psychology: An Introduction to New Concepts in Modern Psychology*, New York: New American Library, 1947.

9. What is the relationship of critical thinking to creativity? We believe that critical thinking includes many of the elements of creativity, but we are presently still working on a model of creative behavior in chemical engineering. An interesting general analysis, however, has recently been offered in a study of forty winners of the prestigious MacArthur Foundation Fellowships. In *Uncommon Genius* (New York: Viking, 1990), Denise Shekerjian offers the following principles of creative thought: discover your talent; commit to it; welcome risk and uncertainty; develop a sense of courage; avoid hasty solutions; know yourself; establish conditions in which you work well; abandon excuses for getting started; acknowledge that creativity is obtained through hard work, not through inspiration; invest your work with a sense of purpose and vision; be prepared to shift your perspective, both intellectually and physically; be prepared to sustain intense periods of concentration and drive; encourage luck; align instinct and judgment; be prepared for rejection and isolation; build resiliency; and, no matter what, remember that you are doing your work simply for the love of it.

Assignment for Sophomores: The Facts

Background

Your ability to think critically and present information lucidly is crucial to success in any field. In this exercise you will be asked to review, analyze, and explain a series of facts on a specific topic. The exercise is not designed to test your knowledge; rather it will allow your instructor to find out how well you analyze, reason from, and write about new information. You should be able to perform the following task if you are willing to think critically about the information provided.

Introduction

Facts by themselves have little meaning; it is only when we begin to see relationships among facts that patterns begin to emerge. For the task you are about to perform, you are to analyze a list of facts and explore the relationships you find among the facts. You are to then present your conclusions about these relationships in a series of organized paragraphs.

The task is organized into four parts:

Part 1 asks you to draw three conclusions.

Part 2 asks you to select one of these conclusions, tell how you arrived at that conclusion, and provide evidence for that conclusion.

Part 3 asks you to tell how you would check the accuracy of the conclusion you reached in Part 2.

Part 4 asks you to tell why we should be cautious about coming to such conclusions.

Before you begin Parts 1-4, study the 21 facts provided below. They are derived from a survey of 4,759 readers of *Chemical Engineering*. (For the survey, see the September 18, 1983 issue, pp. 48-60.) As you examine the data, think about the following: Which items can be connected meaningfully? Which conclusions can be drawn from this data? How should the accuracy of the data being used be checked? Why should caution be used when using data to draw conclusions?

(more)

Assignment for Sophomores:
The Facts

1. 73% of engineers believe that they were well prepared for their first professional jobs. This percentage is about the same for managers, non-managers, older engineers, and younger ones.

2. 96% of chemical engineers report that classroom examples drawn from industrial experience are desirable.

3. 36% of chemical engineers report that they have found differential equations helpful in their careers.

4. 31% of chemical engineers studied oral and/or written communications in college.

5. 23% of engineers are involved in plant operation.

6. 51% of older engineers report that they have taken computer programming in college.

7. 29% of chemical engineer report that they have had co-op experience while in college.

8. 94% of young engineers report that they have taken computer programming in college.

9. 40% of engineers report that they have taken computer programming courses after college.

10. 26% of engineers are involved in design and construction.

11. 21% of engineers are involved in research and development.

12. 99% of chemical engineers studied calculus in college.

13. 19% of engineers report that they have taken no college courses in legal, social, and ethical issues.

14. 30% of recently graduated engineers have not studied environmental engineering in college.

(more)

Assignment for Sophomores:
The Facts

15. After graduation, 61% of chemical engineers report that they found mass/energy courses most helpful.

16. 93% of engineers who worked during school in a co-op program thought the experience more worthwhile than spending the time working on academic courses.

17. 72% of chemical engineers report that they have found legal, social, and ethical courses helpful in their careers.

18. 68% of chemical engineers report that they have found process chemistry helpful in their careers.

19. 67% of chemical engineers report that their professors drew examples from industrial experience while in college.

20. 1.1% of engineers are involved in teaching.

21. 87% of chemical engineers report that they have found oral and written communications courses most helpful in their careers.

(more)

Assignment for Sophomores: The Facts

Part 1

Part 1 will help you connect the information provided in the 21 facts. After carefully reviewing the information, think of ways that the facts are connected to one another. Now, draw three conclusions from these facts, explain your reasons for drawing each conclusion, and specify the facts that support each conclusion.

Part 2

Of the three conclusion you have presented, select the one that you believe is most strongly supported by the 21 facts. Present the conclusion, explain your reasons for drawing this conclusion, and specify the facts that inform that conclusion.

Part 3

What additional information would you need to check the accuracy of the conclusion you drew in Part 2? Where and how would you find such information? Identify the conclusion you reached in Part 2, explain your reasons for wanting to investigate the accuracy of the information, and describe a plan for how you would go about checking the information.

Part 4

Why should we be cautious in drawing conclusions on the basis of isolated facts? Present your answer to this question using specific examples. Your examples may be drawn from this task as well as from other sources.

Assignment for Juniors: Surveying Your Progress in the Chemical Engineering Curriculum

Step 1

The matrix below identifies prerequisites for the chemical engineering laboratory. To survey the quality of your preparation for your senior laboratory course, evaluate yourself on the following scale:

4 - I feel that my abilities are *superior* in this area.

3 - I feel that my abilities are *competent* in this area.

2 - I feel that my abilities are *somewhat competent* in this area.

1 - I feel that I *lack competence* in this area.

Subject	Score			
Chemical Enginerring				
chemical process principles:				
—material and energy balances	4	3	2	1
thermodynamics:				
—generalized methods for handling P-V-T relations	4	3	2	1
—thermodynamic properties of fluids, batch, and flow processes	4	3	2	1
—treatment of compressors, heat engines, refrigeration, phase equlibria, chemical reators	4	3	2	1
—energy utilization	4	3	2	1
transport operations:				
—molecular and turbulent transport of momentum, heat and mass with applications to the design of chemical process equipment	4	3	2	1

(more)

27

Assignment for Juniors: Surveying Your Progress in the Chemical Engineering Curriculum

Subject	Score			
Chemical Engineering (cont.)				
reation kinetics:				
—mechanisms and kinetics of homogeneous chemical reactions in batch and flow reactors	4	3	2	1
—application of kinetics to isothermal and nonisothermal reaction design	4	3	2	1
Chemistry				
physical chemistry:				
—properties of ideal and non-ideal gasses, liquids, and solutions	4	3	2	1
—thermochemistry	4	3	2	1
—thermodynamics	4	3	2	1
—phase rule	4	3	2	1
—phase equilibria	4	3	2	1
—homogeneous and heterogeneous chemical equilibria	4	3	2	1
—ionic equilibria	4	3	2	1
—electrochemistry	4	3	2	1
—kinetic theory of gasses	4	3	2	1
—transport phenomena	4	3	2	1
—kinetics	4	3	2	1
—irreversible processes	4	3	2	1
—physical chemical instrumentation methods	4	3	2	1

(more)

Assignment for Juniors: Surveying Your Progress in the Chemical Engineering Curriculum

Subject		Score		
Chemistry (cont.)				
organic chemistry:				
—preparation of the various classes of organic compounds	4	3	2	1
—commercial utilization of coal and petroleum	4	3	2	1
Mathematics				
calculus:				
—theory and techniques of differentiation and integration	4	3	2	1
—differentiation and integration of inverse trigonometric, exponential, and logarithmic functions	4	3	2	1
—infinite series	4	3	2	1
—applications of the definite integral to physical problems	4	3	2	1
—partial differentiation	4	3	2	1
—multiple integrals	4	3	2	1
—vectors	4	3	2	1
—Fourier series	4	3	2	1
—the expansion of functions	4	3	2	1
—statistics	4	3	2	1
differential equations:				
—solution of ordinary differential equations	4	3	2	1

(more)

Assignment for Juniors: Surveying Your Progress in the Chemical Engineering Curriculum

Subject	Score			
Mathematics (cont.)				
—Laplace transforms	4	3	2	1
—numerical and series solutions	4	3	2	1
Physics				
—simple harmonic motion	4	3	2	1
—wave motion	4	3	2	1
—geometric and physical optics	4	3	2	1
—electricity and magnetism	4	3	2	1
Humanities and Social Sciences				
rhetoric:				
—composition	4	3	2	1
—technical communication	4	3	2	1
history:				
—world history	4	3	2	1
—American history	4	3	2	1
social sciences:				
—technology and social science	4	3	2	1
—policy studies	4	3	2	1
—microeconomics	4	3	2	1
—macroeconomics	4	3	2	1

Assignment for Juniors: Surveying Your Progress in the Chemical Engineering Curriculum

Subject	Score			
Humanities and Social Sciences (cont.)				
management practices:				
—organizational motivation and morale	4	3	2	1
—scientific management and human relations	4	3	2	1
—influence of industrial engineering	4	3	2	1

Step 2

Make a list of the subjects you scored either 2 (indicating that you were somewhat competent) or 1 (indicating that you lacked competence). Go over these areas with your academic advisor and plan strategies for correcting these problematic areas.

Assignment for Sophomores, Juniors, and Seniors: Do Other Instructors Use Critical Thinking Models?

If you recall, in the Preface we stated that this textbook was about experimentation. In this assignment, we would like you to perform an experiment to determine if instructors of various classes have instructional models that are similar to the one you have been using in this chapter.

Below is a blank critical thinking model. Select one of your courses in chemical engineering and one of your courses in social science and interview both of your instructors to determine the instructional model that is used.

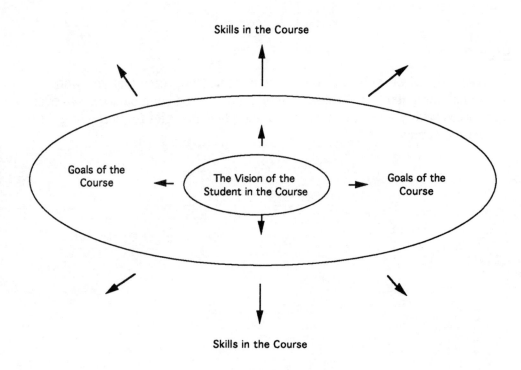

After assembling the blank model with your instructor, analyze how that model may be compared to that used in this chapter. Speculate on why there are similarities and differences.

What is Thinking?

"So what is thinking? Briefly, it is the creation of mental representations of what *is not* in the environment. Seeing a green wall is not thinking; imagining what that wall would be like if it were repainted blue is. Noting that a patient is yellow is not thinking; hypothesizing that the patient may suffer from liver disease or damage is . . . ."

" To oversimplify, there are basically two types of thought processes: automatic and controlled. The terms themselves imply the difference. Pure association is the simplest type of automatic thinking. Something in the environment 'brings an idea to mind.' Or one idea suggests another, or a memory. As the English philosopher John Locke (1632-1706) pointed out, much of our thinking is associational."

"At the other extreme is controlled thought—in which we deliberately hypothesize a class of objects or experiences and then view our experiences in terms of these hypothetical possibilities. Controlled thought is 'what if' thinking. The French psychologist Jean Piaget (1896-1980) defined such thinking as 'formal,' in which reality is viewed as secondary to possibility. Such formal thought is only one type of controlled thinking. Other types include visual imagination, creation, and scenario building."

Robyn M. Dawes, *Rational Choice in an Uncertain World* (1988)

"Because we do not come to our experience with a blank slate for a mind, because our thinking is already, at any given moment, moving in a direction, because we can form new ideas, beliefs, and patterns of thought only through the scaffolding of our previously formed thought, it is essential that we learn to think critically in environments in which a variety of competing ideas are taken seriously."

—Richard Paul, *Critical Thinking* (1990)

2 Interpreting the History of Chemical Engineering

Gordon Lewandowski
Professor of Chemical Engineering

"Those who cannot remember the past are condemned to repeat it."

George Santayana

Preview

In Chapter 2, you'll have a chance to think about

- the history of chemical engineering
- a systematic method of historical analysis

"For I have several boxes in my memory in which I will keep them all very safe, there shall not one of them be lost."

Isaac Walton, *The Compleat Angler*

A Sketch of Chemical Engineering

Engineers are not applied scientists. Engineers use applied science—and anything else they can—to solve problems. By their very nature, the best engineers are critical thinkers.

Engineering and science have very different pedigrees. Science derives from philosophy—from Aristotle (384–322 B.C.), Descartes (1596–1650), and Newton (1642–1727). Conversely, engineering derives from craft skills. The civil engineers who built the Roman aqueducts were master masons who began their careers by cutting stone. The great Venetian architect Mauro Coducci (1440–1504) was a stone mason from Bergamo. James Watt (1736–1819), who invented an improved steam engine, was an instrument maker from Glasgow.

The early, yet powerful, need for practical application is expressed in Marcus Vitrivius Pollio's *De Architectura*, written in approximately 15 B.C. In the last section of the work, Vitrivius includes a description of the machines of his day: hoisting gear, derricks, pulleys, and pumps. There is no mention of the theory of mechanics. "Those who rely on theories are obviously hunting the shadow, not the substance."[1] Engineers derive from a very long line of people who had little knowledge of, or interest in, theory. They did, however, have a strong need to solve problems of a physical—as opposed to a metaphysical—nature.

Formal education of engineers did not take place until the beginning of the 19th century, as a result of military needs, such as the manufacture of cannon and the construction of fortifications. France was the pioneer in this regard. The formation of the Ecole Polytechnique in 1794 by the French geometer Gaspard Monge is of enormous importance in that "here, for the first time, engineers were being trained in fundamental science by means of laboratories and experiments".[2] The United States Military Academy at West Point, founded in 1802, was America's first engineering school, with a curriculum modelled after that of the Ecole Polytechnique. Later on, the increasing needs of an industrial society, coupled with the growing complexity of technological applications, led to the founding of private engineering colleges, such as Rensselaer Polytechnic Institute in 1824, and the Massachusetts Institute of Technology in 1861.

This process of engineering education again points to the different heritages and pedigrees of engineering and science.[3] Existing universities were often hostile to the formal development of a profession (engineering) in which doers were more valued than thinkers (theorists), and in which approximations were not only tolerated, but praised, if they could help solve problems of a practical nature. By contrast, science usually demands orthodox rigor.

Nevertheless, the replacement of apprenticeship in craft skills with formal education at engineering colleges also led to an increase in the application of scientific methods to engineering practice. Particular observations were followed by attempts to generalize those observations to other problems. Engineering may be defined as the organization and extrapolation of experience. Mathematics is always used, and science often used, as the tools in this process.

The engineering profession still suffers today from the stigma attached to craft skills and problem solving (rather than "higher order" intellectual pursuits involving theoretical and philosophical considerations). This situation is compounded by the fact that engineers only need a bachelor's degree to be considered practicing professionals, rather than a doctorate as in the sciences. Unfortunately, this has often created a split between the educational hierarchy in engineering (which in many universities aspires to be science) and the practice of engineering, which is highly pragmatic.

Humanity has greatly benefitted from the profession of engineering. The engineering practice of separating our drinking water supplies from our sewage has probably saved more human life than all of the medical doctors that have ever practiced. There are also the enormous contributions by chemical engineers to the production of pharmaceuticals, pesticides, and fertilizers (without which we could not feed nearly 5 billion people). It is important for engineers to value their profession, without historical apology.

Until the second half of the 19th century, there were basically only two formal engineering disciplines (what would be termed today mechanical and civil engineering). Thereafter, first electrical, and finally chemical engineering came into being.

As an emerging profession, chemical engineering had difficulty distinguishing itself from both chemistry and mechanical engineering. (Indeed, to this day in France and Germany, industrial chemists often do the work that in the U.S. is done by chemical engineers). Nevertheless, the young discipline began to gain credibility, partially by building a curriculum different from that of chemistry and mechanical engineering.

Programs in chemical engineering had been formed at the Massachusetts Institute of Technology as early as 1888.[4] By 1908 there were over a dozen such programs. By 1905, Richard K. Mead, founder of a new periodical named *The Chemical Engineer*, proposed an American society of chemical engineers. In 1908, the discipline formed itself as the American Institute of Chemical Engineering.

There remained, however, a lack of educational uniformity. The catalyst that gave the discipline both its foundation and its force came in 1915 in the form of Arthur D. Little's concept of "unit operations." Founder of an industrial consulting firm, Little wrote the following to the president of MIT:[5]

> Any chemical process, on whatever scale conducted, may be resolved into a coordinate series of what may be termed 'Unit Operations,' as pulverizing, dyeing, roasting, crystallizing, filtering, evaporation, electrolyzing, and so on. The number of these basic unit operations is not large and relatively few of them are involved in any particular process. The complexity of chemical engineering results from the variety of conditions as to temperature, pressure, etc., under which the unit operation must be carried out in different processes, and from the limitations as to material of construction and design of apparatus imposed by the physical and chemical character of the reacting substances.

This approach has had a profound impact on the nature of our profession, and sets us apart from other engineering professions. It is the engineering equivalent of the assembly line. Instead of skills being applied to each new problem (as in civil engineering, in which each new building or bridge is unique), chemical processes are constructed from a series of prefabricated parts called "unit operations." As a result, chemical engineering does not involve applying skills to a given process—such as soap production or petroleum separation—as industrial chemistry might. Rather, chemical engineering stresses the application of skills to unit operations, from which an infinite variety of chemical processes may be constructed.

To return to our comparison, a civil engineer will view a secondary waste treatment process as a unique construct requiring unique engineering tools, but a chemical engineer will view the same process as a series of familiar "unit operations" (precipitation, followed by biochemical reaction, followed once again by precipitation), each of which is found in many other processes, such as pharmaceutical production. Thus, the chemical engineer will be able to make important connections and inferences between one process and another. Unit operations remains a core principle of chemical engineering education.

An important result of the study of unit operations is its impact on scale-up. This is an essential chemical engineering function that takes the results obtained by chemists at a laboratory (or "bench") scale, and transfers that process to commercial practice at a very large scale. A commercial plant for producing (for example) aspirin, is not simply a very large version of the equipment and process steps used by a laboratory chemist. For one thing, the chemist will use glassware, whereas the commercial plant will be built entirely of stainless steel. The chemist will employ a batch process, filling and emptying vessels sequentially, whereas the commercial facility will continuously produce aspirin tablets from a continuous input of raw materials. "Scale-up" involves engineering science and problem solving skills. Its attendant development costs are orders of magnitude greater than the original research expenditures of the laboratory chemist. A very interesting description of the history of chemical process development and scale-up can be found in Science and Corporate Strategy: DuPont R&D, 1902–1980.[6]

The history of trends in modern chemical engineering may be seen in its textbooks, as Olaf A. Hougen has shown.[7] The principle of unit operations, for example, was introduced in *Principles of Chemical Engineering* (1923) by William H. Walker, Warren K. Lewis, and William H. McAdams. From 1925 to 1935, the calculation of material and energy balances was captured for students by O. A. Houghen and K. M Watson in *Industrial Chemical Calculations* (1931), and work in thermodynamics was explained for the classroom in H. C. Weber's *Thermodynamics for Chemical Engineers* (1939). An important advancement in the profession came in 1960 with the publication of *Transport Phenomena* by R. B. Bird, W. E. Stewart, and E. N. Lightfoot. This text took the idea of unit operations one step further by showing that the mathematical skills used to treat each of these operational processes have an underlying similarity. This concept allows even greater generalization of chemical engineering principles to such fields as human physiology and microbial processes.

For the future, there are a number of opportunities for chemical engineering in its application to interdisciplinary fields, such as hazardous waste treatment, biomedical engineering, biochemical engineering, and materials processing.[8] Our training in chemistry, as well as physics, presents these opportunities to us.

The essence of engineering has always been problem solving. It is in the spirit of developing additional problem solving skills that the present textbook is dedicated. You are being asked to extend your engineering tools to include critical thinking, ethics, and communication.

The Use of Historical Analysis

This brief sketch of the history of chemical engineering illustrates its importance in western culture. As a field of exploration, it helped to shape the cultures that supported it.

However, it is quite difficult for any undergraduate student to command a really detailed and thorough history of chemical engineering. The developments are far too vast, the details far too subtle. Nevertheless, an historical orientation is essential if students are to think critically about the projects they encounter in the university and in the workplace. All chemical engineering exists in complex contexts. The more you know about these contexts, the better you will be able to proceed.

What does such an orientation involve? The best way to understand what we mean by an historical orientation to problem solving is for you to work through an example. Consider the case below:

August, 1941

In early September of 1940, Hitler's Luftwaffe began fire bombing England. At Oxford, Professor Howard W. Florey and his research team began investigating better treatment methods for burns. Resurrecting Sir Alexander Fleming's discovery of penicillin, the Oxford group made enough penicillin for clinical trial and quickly became convinced of its miraculous results.

Early recovery of penicillin, however, was very poor. Growing penicillin in ceramic pans, the Oxford group could produce only 3 units per milliliter. How could enough penicillin be produced to aid the Allied war effort?

How might you interpret this historical episode? To produce a systematic analysis, we suggest that you use a series of useful questions which are sometimes called a heuristic.

Systematic Historical Inquiry: A Method

Richard E. Young, Alton L. Becker, and Kenneth L Pike have developed a powerful method (heuristic) of providing "specific plans for analyzing and searching which focus attention, guide reason, stimulate memory and encourage attention."[9] These researchers argue that any event, such as the history of penicillin production, can be analyzed by means of three cognitive activities:

Contrast: This type of analysis allows you to see the features that make an event different from other similar events.

Variation: This type of analysis allows you to see the different kinds of features within an event.

Distribution. This type of analysis allows you to see how an event may be understood within a larger context.

Each of these three cognitive activities may also be analyzed in three ways:

A particle view will allow you to see an event as isolated.

A wave view will allow you to see an event as dynamically related to other events.

A field view will allow you to see an event as part of a larger network of relationships.

To visualize this system of analysis, Young, Becker, and Pike offer a chart similar to the one shown in Figure 2.1.

	Contrast	Variation	Distribution
Particle	1. View the event as an isolated problem.	4. View the range of the event.	7. View the event as part of a larger classification.
Wave	2. View the event as a dynamic incident.	5. View the event as a dynamic process.	8. View the event as a part of a dynamic context.
Field	3. View the event as a system.	6. View the event as a system of many dimensions.	9. View the event as a system within a larger system.

Figure 2.1 A System for Historical Analyses

Note that there are nine cells in the chart. Let's now apply this analytic system to the example of penicillin production by viewing our case in the particular way that each of these nine cells indicates:

1. What is the nature of the scale-up problem faced by the Allies?

2. Penicillin production may be contrasted with other chemical engineering practices in the early 1940s. How are the problems faced by the Allies different from other chemical engineering problems faced during that period? For example, were the problems of penicillin production different from other fermentation processes?

3. Using this frame of reference, you may contrast the problems faced by Allies with the kinds of problems that chemical engineers face in any situation. At what point did researchers at Oxford turn to industry for help? How is this decision different from other demands for scale-up?

4. In examining penicillin at its discovery, you encounter the saga of Sir Alexander Fleming's discovery. In looking at penicillin production today, you might make a table of the average annual production rate, the treatments of patients per month, and the average annual value of the penicillin.

5. How is the process of penicillin production changing? By looking closely at the table suggested for cell #4, you can pose questions about the nature of the changes in penicillin production.

6. How did the parts of the process of penicillin scale-up differ? How, for example, were the problems of equipment design different when engineers addressed (a) the need for sterile air which could be injected into a submerged culture and (b) the need for greater recovery of the product?

7. How does the field of chemical engineering classify the production of penicillin?

8. Does this field—pharmaceuticals—differ from petroleum? How is the pharmaceutical field changing today?

9. Finally, you might think about the significance of penicillin production within the larger history of chemical engineering.

The Importance of Systematic Problem Solving

Why undertake such a difficult and rigorous process? Young, Becker, and Pike offer three answers.

First, this type of systematic inquiry will help you to retrieve relevant information that you already have in your grasp. Students, we have found, know far more then they think, but they often cannot devise a way to recapture that knowledge. The system presented above will allow you to avoid drawing a blank.

Second, the system points the way to new information that you must seek. A bibliographic search would provide some of the necessary sources.

Finally, the system prepares you to create your own frame of reference. After you have surveyed what you already know and determined what you must investigate, you will still need a way to organize your material. By referring to the nine-cell historical analysis, you will probably be drawn to several particularly intriguing questions. These questions might prove to be the key ordering principle that will allow you to understand the significance of a particular chemical engineering process.

We offer one more reason to this list: historical understanding is essential to successful chemical engineering practice. Only by understanding the problems of the past can we understand the problems of the present. In reading *The History of Penicillin Production*[10], for instance, you would no doubt discover the sense of pride that the engineers felt in their collaborative effort. Wrote E. D. Coghill of his experience, "The penicillin battle was not won by any one group, but by a tremendous cooperative effort. University laboratories, government laboratories, and industrial laboratories all worked together on it." It is that sense of collaboration and pride that you can carry with you into the chemical engineering laboratory as well as into your future professional development.

Notes and References

1. Quoted in James Kip Finch, *The Story of Engineering*, New York: Anchor Books, 1960, p. 58.

2. W. H. G. Armytage, *A Social History of Engineering*, New York: Pitman Publishing Corporation, 1961.

3. Edwin Layton, "Mirror-Image Twins: The Communities of Science and Technology in 19th–Century America," *Technology and Culture*, (October 1971), pp. 562–580.

4. Harold C. Weber, *The Improbable Achievement: Chemical Engineering at MIT*, Cambridge, MA: MIT Press, 1979.

5. Terry S. Reynolds, *75 Years of Progress: A History of the American Institute of Chemical Engineers*, New York: AIChE, 1983, p. 10.

6. Olaf A. Hougen, "Seven Decades of Chemical Engineering." *Chemical Engineering Progress*, (January 1977), pp. 89–104.

7. David A. Hounshell and John K. Smith, Jr., *Science and Corporate Strategy: DuPont R&D, 1902–1980*, New York: Cambridge University Press, 1988.

8. For more on the future of chemical engineering, see: Edward G. Jefferson, "The Emergence of Chemical Engineering as a Multidiscipline," *Chemical Engineering Progress*, (January 1988), pp. 21–23; Robert M. White, "Technological Competitiveness and Chemical Engineering," *Chemical Engineering Progress*, (January 1988), pp. 24–26; Isaac Asimov, "The Future of Chemical Engineering," *Chemical Engineering Progress*, (January 1988), pp. 43–49; and National Research Council, *Frontiers in Chemical Engineering*, Washington, DC: National Academy Press, 1988.

9. For a textbook devoted to this method of analysis (called tagmemic invention), see Richard E. Young, Alton L Becker, and Kenneth L. Pike, *Rhetoric: Discovery and Change*, New York: Harcourt, Brace, Jovanovich, 1970. See also Richard E. Young, "Invention: A Topographical Survey" in *Teaching Composition: 10 Bibliographic Essays*, ed. Gary Tate, Fort Worth: Texas Christian University Press, 1976.

10. A. L. Elder, ed., *The History of Penicillin Production*, New York: American Institute of Chemical Engineers, 1970.

11. Kenneth Burke, *A Grammar of Motives*. New York: Prentice Hall, 1945.

Assignment for Sophomores and Juniors: Writing About the History of Chemical Engineering

In this assignment we want you to write a research paper on some historical aspect of the field of chemical engineering. To help you narrow the topic, we've selected significant dates below. You may select any one of these topics as the focus of your paper:

1908—The American Institute of Chemical Engineers (AIChE) is founded.

1912—The AIChE adopts its first code of ethics.

1915—A. D. Little coins the term "unit operations."

1923—W. H. Walker, W. K. Lewis, and W. H. McAdams publish *Principles of Chemical Engineering*.

1960—R. B. Bird, W. E. Stewart, and E. N. Lightfoot publish *Transport Phenomena*.

1962—Rachel Carson publishes *Silent Spring*.

1987—Edward G. Jefferson, former CEO of DuPont, tells an audience in New York that "there will be times when understanding of political technology will be as important [to our students] as knowledge of thermodynamics or chemical kinetics."

To begin, select any one historical moment and employ Young, Becker, and Pike's nine-part heuristic to help you ask questions systematically. As you use the heuristic, think about the least interesting question you asked. Why was it uninteresting? On the other hand, what was the most interesting question you asked. Why did it intrigue you?

Now that you have narrowed your topic, you should begin your research. To help you get started on a review of sources about the history and development of chemical engineering, review the sources Gordon Lewandowski used in his essay. In addition, review the discussion of the historical paper in Chapter 10. You'll also find there a sample student paper.

The kind of historical research you will perform in this paper is very important to your solid grasp of the present and future development of chemical engineering.

Unit Operations: Origins of The Principle and Its Relationship to National Boundaries

It was an idea of genius. Before it, chemical engineers had a difficult time even marking the boundaries of their work. After it, the profession achieved a unique identity that made it the leader of the big four engineering fields.

The concept of unit operations is a familiar one that A. D. Little first introduced in 1915 and reiterated in 1922: "Chemical Engineering as a science . . . is not a composite of chemistry and mechanical and civil engineering, but a science of itself, the basis of which is those unit operations which in their proper sequence and coordination constitute a chemical process as conducted on the industrial scale."

Did A. D. Little invent this concept? Historians tell us that the concept was in the air, created by a kind of *zeitgeist* that emphasized the importance of systems.

Then where else can we find the concept? By 1911, Frederick W. Taylor had published *The Principles of Scientific Management*, a book describing the advantages of detailed job analysis. Known for their time and motion studies, Taylor's disciples wished to determine scientifically the nature of any industrial task. This kind of analysis appears to have influenced Taylor when he proposed the concept of breaking down and analyzing chemical operations. As Terry S. Reynolds has argued, the concept of unit operations not only sharply distinguished chemical engineers from analytic chemist but also allowed the profession to move "cognitively" away from chemistry. Hence, unit operations may be understood itself to be a form of critical thinking.

But was this the pattern in other countries? Jean-Claude Guedon argues that it was not. "The organizing principles guiding European chemical industries," he writes," evolved with a great deal of specificity coinciding with national boundaries. As a result, each country came to a kind of modus vivendi which maintained itself as long as international competition remained on a commercial plane."

When the First World War broke out, however, things changed. Germany had established a unique method of training chemical engineers by having chemists circulate through the different chemical divisions of a plant until they were familiar with all the aspects of chemical activities in that plant. Then the individual could work in that area that best suited his

(more)

Sidebar

abilities. The field of chemistry, the Germans felt, was too large to master, so what was needed was a division of labor created by the apprentice process described above. Because France and England had no such underpinnings for the chemical industries, they were at a loss as World War I approached. While England was able to close the gap eventually by adopting America's unit operations model, France was unable to conceptualize a plan for chemical engineering.

What may we conclude from this slice of history? First, it is clear that individuals are influenced by the periods in which they live. Little's ideas are related to Taylor's in a way that illustrates that both men were responding to an expressed need for systematization that was characteristic of the period in which they lived. Second, it is clear that individuals are influenced by the geography in which they find themselves. Guedon argues that what we see if we compare the United States, Great Britain, France, and Germany is a case of "national technological styles."

As we approach the millennium, it may well be worth our while to speculate a bit on how the rise of a multinational economy will shape the ways in which we conceptualize the challenges before us.

Sources:

Jean-Claude Guedon. "Conceptual and Institutional Obstacles to the
        Emergence of Unit Operations in Europe." In *History of Chemical*
        Engineering. William F. Furter, ed. Washington, D. C. : American
        Chemical Society, 1980. pp. 45-75.
Terry S. Reynolds. "Defining Professional Boundaries: Chemical
        Engineering in the Early 20th Century." *Technology and Culture*
        27 (1986): 694-716.

3 Working in the Laboratory

"Everything which can be encompassed by man's knowledge is linked in the same way, and that provided only that one abstains from accepting any for true which is not true, and that one always keeps the right order for one thing to be deduced from that which precedes it, there can be nothing so distant that one does not reach it eventually, or so hidden that one cannot discover it."

Rene Descartes, *Discourse on Method*

Preview

In Chapter 3, you'll have a chance to think about

- the way that research in chemical engineering is conducted

- a detailed laboratory research plan

- guidelines for planning and safety

- characteristic problems in laboratory performance

- planning laboratory experiments

- the importance of collaboration

- the value of criteria-based assessment of your work

"For Angling may be said to be so like Mathematics, that it can ne'r be fully learnt; at least not so fully, but that there wil stil be more new experiments left for the trial of other men that succeed us."

Isaac Walton, *The Compleat Angler*

The Significance of the Laboratory as a Test Bed for Chemical Engineering

Why place so much emphasis on the laboratory? A lesson from history will show us why the laboratory is central to chemical engineering.

As David A. Hounshell and John Kenly Smith illustrate in their monumental history of science and corporate strategy at Du Pont, modern industrial and research development laboratories "did not emerge, full blown from the minds of executives at such firms as General Electric, Du Pont, Eastman Kodak, and American Telephone and Telegraph."[1] Perhaps the "invention factory" of Thomas A. Edison offers the precedent for a Research and Design (R & D) laboratory. In 1876 Edison established a laboratory at Menlo Part, New Jersey, with the promise that the laboratory would produce "a minor invention every ten days and a big thing every six months or so."[2]

In Menlo Park, Edison built the best-equipped laboratory in the United States. He made good his claims: research on electric lighting was begun in 1878; by 1881 Edison was ready to construct his first American commercial central station. (An insight into Edison's character may be gained from noting that he named the first generator "Jumbo" after the elephant brought to America by P. T. Barnum.[3])

It is no wonder that U.S. corporations such as DuPont turned to this kind of formally organized research. This type of laboratory, then flourishing in Germany, was characterized by what Heinrich Caro, director of Badische Anilin-und-Soda Fabrik (BASF), referred to as *wissenchaftliche Massenarbeit*: "massive scientific teamwork." Massive, corporate-driven research efforts came to be adopted by large corporations throughout the United States. Under such influence, between 1902 and 1903 Du Pont founded its first research laboratories. The American chemical development industry had begun, and the laboratory was established as its unique, creative arena—a place where ideas were born, developed, and tested.

What are some important things to appreciate and respect about this discipline's arena—both in your school and in the corporate world?

First, the chemical engineering laboratory is a realistic environment. While your classroom experience in, for example, reaction kinetics may seem to be isolated, the application of kinetics in an experiment involving a non-isothermal, non-adiabatic reactor is realistic. The laboratory, both within and beyond the university, is filled with cases in which theory is aligned with practice.

Second, research is collaborative. Again recalling your kinetics class, you will remember that assigned problems were solved by each of you working in a setting of your choice (late at night, radios blaring). On the contrary, in the laboratory, the experiments—driven by *wissenchaftliche Massenarbeit*—are conducted by teams. In the laboratory, our sense of self-reliance must now be brought into alignment with others' strengths and weaknesses. Who is theoretically able? Technically and mechanically proficient? Who is the best at oral presentations? At writing various kinds of reports? Success in the laboratory is critically linked to the laboratory team's ability to realize and utilize the skills of its members. Hence, your experiences in the laboratory as an undergraduate are especially crucial if you are to be ready for the world you will face at corporations, such as DuPont, upon graduation.

Third, laboratories are agenda-driven. Your course work and your professors demand the completion of specific tasks by specific dates, and your educational progress depends on the systematic mastery of certain laboratory techniques and experiments. Similarly, laboratories in the corporate world are driven by corporate decisions about research agendas, production dates, and funding.

Fourth, successful laboratory work is dependent upon demonstrable outcomes. Laboratories are the discipline's arenas of proof. Your university laboratory is where textbook theories and facts prove themselves and where you prove yourself in command of the practical aspects of science. The corporate laboratory is where chemical engineers demonstrate the corporation's return on its investment—the process or product that will prove profitable in a commercial market.

Fifth, the successful laboratory experience depends upon the effective communication of plans and outcomes. Remember, the laboratory is collaborative and agenda-driven; those involved in a collaborative effort are dependent upon the effective communication of the agenda. And, since laboratory work involves demonstrable outcomes, students and chemical engineers must be able to communicate their findings effectively. Students need to communicate their outcomes to lab collaborators and instructors; chemical engineers need to communicate their outcomes in order to obtain funding for proposed research or to receive recognition for completed research.

We hold, therefore, that the chemical engineering laboratory is perhaps the most important part of your undergraduate education. How else might you better understand the laboratory as a reflection of the field? To extend and deepen your background, we will now present an organizational system for chemical engineering.

Part I: A Paradigm for Chemical Engineering in the Laboratory

The term paradigm was popularized by science historian Thomas Kuhn.[4] A paradigm may be defined as the way that a discipline conducts its research. What is the paradigm for chemical engineering? We have presented a way for you to understand the comprehensiveness of the field in Figure 3.1.

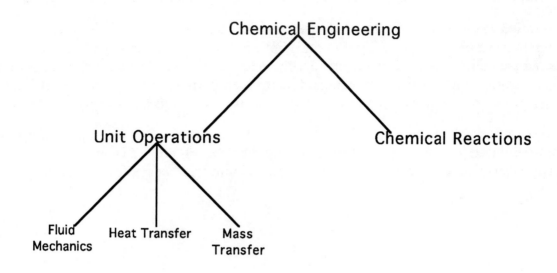

Figure 3.1 An Organizational Paradigm for the Field of Chemical Engineering

As Figure 3.1 reveals, chemical engineering may be divided into two fields: Unit Operations and Chemical Reactions. Unit Operations generally deals with the physical processing of materials and streams. Chemical Reactions, on the other hand, concerns all chemical conversions of species.

The field of Unit Operations may be subdivided into three major branches: Fluid Mechanics, Heat Transfer, and Mass Transfer. These divisions are based fundamentally on balances of momentum, energy, and material, respectively. A partial listing of Unit Operations includes: pipe flow, mixing, packed towers, separations (e.g. distillation, absorption, adsorption, extraction), drying, humidification, evaporation, radiation, and heating. The field of Chemical Reactions, on the other hand, employs the principles of chemical kinetics and equilibrium to engineer organic and inorganic reactions for desired products.

Vastly different chemical engineering plants can be characterized as an integrated grouping of certain common Unit Operations and Chemical Reactors. For example, a bubble cap column processing crude oil in a petroleum refinery and a sieve tray column processing halogenated methanes in a fire retardant manufacturing plant are each separations (distillation) in which the stage operations, phase equilibrium, contacting, and heat transfer are employed.

It is imperative that you bear in mind that this model of the chemical engineering discipline is designed to provide a degree of order to what might otherwise seem a hodge-podge of topics. However, this model should not suggest that an experiment listed under a specific branch is devoid of any consideration of the other branches. For example, the shell-and-tube heat exchanger is evaluated for heat balance and heat transfer coefficients. This categorization, however, does not mean that fluid mechanical pressure drop is not occurring. You are just not considering it in your experiment. Similarly, species separation is the primary consideration of the binary distillation experiment; however, the heat balance on the column must be taken into consideration. Reaction engineering, on the other hand, cannot take place without consideration of heat effects.

In addition to cross-overs in the field, chemical engineering also includes other important topics without which it could not exist as a discipline. The most important of these is thermodynamics. As presently taught, thermodynamics includes phase equilibrium, reaction equilibrium, work/heat relationships, pure component and mixture properties, and other topics. Because thermodynamics crosses so many boundaries, it does not fit easily into the paradigm of Figure 3.1.

In turning now to your course, you can see how the experiments in your university's laboratory may be placed into the organizational paradigm below. Figure 3.2 is an example of a university's chemical engineering experiments placed into the model shown in Figure 3.1.

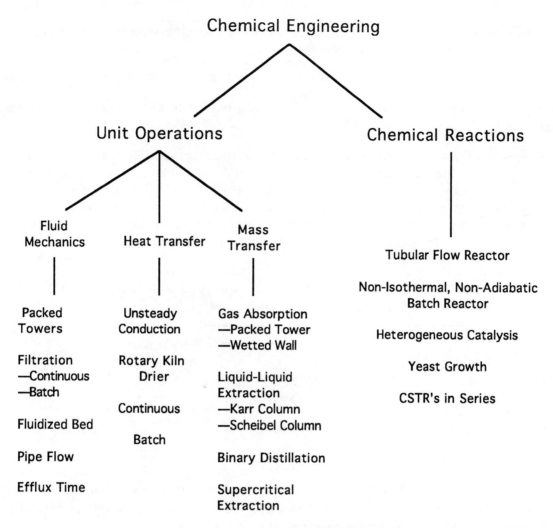

Figure 3.2 Selected Experiments as Examples of the Organizational Paradigm for Chemical Engineering

We turn now to an elaboration of an experiment in the model found under mass transfer: binary distillation.

A Sample Experimental Description: Mass Transfer

In the process of writing this textbook, we surveyed the unit operations courses in twenty leading universities in the country. The description of an experiment often contained these four sections:

Introduction: The primary objectives of the experiment are directly stated here. The overall experimental approach to meet the objectives is also provided.

Suggested Procedure: Detailed instructions are provided here for operation of the equipment and collection of the data. Sufficient detail is given so that students can effectively plan and safely execute the experiment. However, this information is more provocative than explicit. In other words, the laboratory instructor has been careful not to write a cookbook set of instructions, thereby providing so much detail that students' imagination and creativity would have no place.

Useful Data: Selected information, such as equipment specifications that are impractical for students to measure, are provided here.

Useful References: A list of helpful research articles and books is provided.

In our own laboratory, we have instituted an additional section for all our experimental descriptions:

Discussion Topics: Various probing questions and issues are listed to stimulate our students to think about their experiments and data and to guide their analyses. In addition to the rather straightforward questions (e.g., What are the number of theoretical stages in a distillation experiment?), our students are asked to think critically about more subtle issues which usually get to the heart of the experiment (e.g., Why are your tray efficiencies less than unity?) The responses to these questions help our students carefully analyze their experiments.

Below is an example of one of our experimental descriptions. It will help set the context for the research plan we will present in section 3.5.

Figure 3.3 A Sample Description of a Mass Transfer Experiment

MASS TRANSFER EXPERIMENT # 4
CONTINUOUS DISTILLATION OF A BINARY MIXTURE

Introduction

In this experiment, a multi-tray continuous distillation column is evaluated. A key objective is to compare the actual number of stages with the ideal number calculated by the McCabe-Thiele and Ponchon-Savarit methods for the observed separation.

There are two evaluation schemes which are practical within time constraints (see instructor as to which to perform). The first is to operate at two different reflux ratios— one finite and one infinite—with a fixed feed tray location. The second is to operate at two different feed tray locations while fixing a finite reflux ratio.

A mixture of methanol and water is available in a reservoir for separation. The column products are returned to the reservoir for reuse.

Suggested Procedure

1. Power up instrument panel. Set nitrogen gas for temperature recorder purge at 0.5 scfh (cylinder outlet at 20 psig). Set purge flow indicators at 0.5 scfh.

2. Open up house air supply. With regulator, set pressure to instruments at 20 psig.

3. Open up house water supply; set condenser cooling water rotameter to about 90.

4. Choose a feed entry point on the column. Open appropriate valve; close the other two.

5. Turn on cooling water flow to bottoms product cooler. Open valve on reboiler steam condensate line.

(more)

6. Set reflux ratio controller on manual. Set a finite ratio (less than 5:1 — typically ca. 2:1). Open top product panel valve full.

7. Pump in feed liquid into the column, without steam on to reboiler, until liquid overflows reboiler weir and fills bottoms product drain.

8. Flow main steam to reboiler. BEWARE HOT SURFACES!! Set column differential pressure controller to MANUAL. Try to operate with set point of about 11 in. water. Liquid level holdup on trays should be about 2–4 in., as viewed in glass section (1ˢᵗ floor). Increasing differential pressure will increase steam to reboiler, resulting in a higher liquid level on the trays.

9. Set feed rate to about 22 on rotameter, using pump vernier.

10. Set the bottoms product sump level controller so that the valve to bottoms pump is wide open.

11. Begin to pump out bottoms product. Use the bottoms product panel valve to manually control the withdrawal rate to keep the product sump from running dry (ca. 3 on the bottoms rotameter). Use a flashlight and observe sump through reboiler window.

12. To achieve steady state, adjust feed rate such that sum of bottoms and distillate product flows is about equal to feed flow rate. Also, look for a reasonable temperature spread across the column which indicates a good separation.

13. After steady state is achieved, take vapor and liquid samples from each tray, condenser, and reboiler. Use the sampling coolers with external ice water bath. Collect samples in stoppered jars, taking care to minimize evaporation. Use compressed air to blow out the cooling coil between samples.

14. Analyze samples with refractometer operated isothermally at 25°C.

15. Measure the reboiler steam rate by redirecting the condensate rate into a tared vessel.

16. For infinite reflux operation, turn off feed pump and close product withdrawal valves. Set reflux ratio to highest setting.

17. Repeat steps 13–15.

Useful Column Data

Diameter: 6" Sch. 10 s.s. pipe; No. of plates: 10

Interplate Distance: 12"; Weir Height: 1";

Condenser Area: 20 ft² - double pass; Reboiler Area: 10 ft²

Discussion Topics

1. What is the overall column efficiency, based on theoretical and actual stages?

2. How do the results from the McCabe-Thiele and Ponchon-Savarit methods compare? Should they differ?

3. How does Murphree tray efficiency vary along the column? Relate this efficiency to the overall column efficiency?

4. If infinite reflux ratio was performed, is the Fenske equation applicable for this system? (Hint: Is relative volatility constant?)

5. If different feed tray locations were studied, how did this parameter affect the column performance?

Useful References

Foust, A. S. et al., *Principles of Unit Operations*, J. Wiley & Sons, New York (1960).

Holland, C. D., *Fundamentals of Multicomponent Distillation*, McGraw-Hill, New York (1981).

King, C. J., *Separation Processes*, McGraw-Hill, New York (1971).

McCabe, W. L. and Smith, J. C., *Unit Operations of Chemical Engineering*, 3rd. ed., McGraw-Hill, New York (1976).

Treybal, R. E., *Mass-Transfer Operations*, 2nd. ed., McGraw-Hill, New York (1968).

Part II: Working Within the Paradigm—How Research is Conducted in Chemical Engineering

Classroom laboratory experience directly parallels research experience in the field of chemical engineering. In this section we will demonstrate how that parallel manifests itself so that you can more readily understand your lab time as preparatory. We begin with Figure 3.4 below, which outlines the chemical engineering research process.

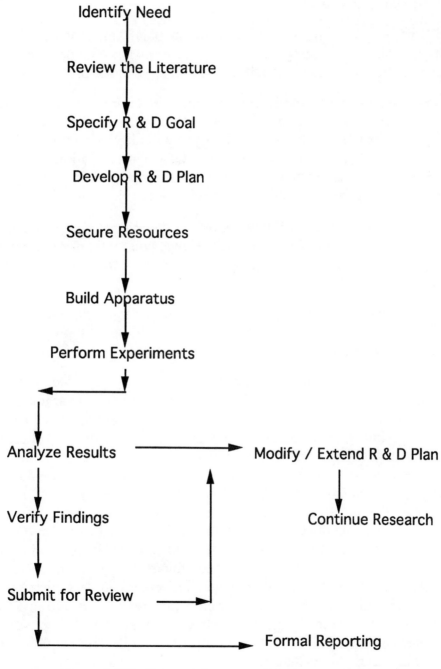

Figure 3.4 The Research Process in Chemical Engineering

In order to start doing research, you must identify a need. There must be some lack of under-standing, some lack of a technology, which must be addressed. Once the need has been identified, you must review the literature in order to establish the "state-of-the-art." Information obtained here will allow you to specify and clarify your research and development (R&D) goal. With a goal in mind, you can begin to formulate a R&D plan consistent with your available or obtainable funding. Here you will decide on appropriate experiments and establish your equipment and personnel needs.

When your planning is completed, you then actually secure your resources, equipment, and personnel. The experimental equipment is then assembled and tested. The experiments are per-formed, and data are collected and recorded.

With data in hand, you begin to critically analyze the data. This process will involve establish-ing the limits of your data and creating a theoretical framework, or model. The interaction be-tween the model predictions and the data will allow you to generate preliminary conclusions.

You must now repeat selected experiments and analyses in order to verify your findings. These early findings might cause you to modify your R&D plan. You now submit your findings for peer review. If you are in industry, you are likely to submit your preliminary report to a supe-rior. In academics, you will submit a report (e.g. scholarly paper) to a journal. The editor will send it for review to at least one recognized expert in your field. Based on those reviews, the editor will recommend the paper for publication, revision, or rejection.

Peer reviews are very important, for they not only provide needed critical comment, but also yield fresh insights into your work. Based on the reviews, you might need to modify your R&D plan. You will also revise your report, accommodating the reviews. You now continue your research to achieve your goal.

How does your chemical engineering laboratory course fit into this research process? With few exceptions, your experiments will fit into every one of the steps shown in Figure 3.4. Let us take the distillation experiment shown in Figure 3.3 as an example.

You identify the need to evaluate the performance of this ten-tray column. Reviewing the literature, you establish how a distillation column operates, which key parameters to monitor, and how to evaluate its performance. This leads you to a research goal which is consistent with the objectives for the experiment: evaluate the column through a measurement of efficiency as deter-mined by theoretical stages. (More about these objectives will be discussed below.) You now develop a plan, guided by the suggested procedure section. Fortunately, you do not have to obtain funding. (Actually, you must pay your tuition!) The column is already assembled, but you must gather the needed supplies and small equipment from the stockroom. You perform the experi-ments, collecting and recording the data in your notebook. You critically analyze the data, setting uncertainty limits. You generate a theoretical model, and compare its predictions to your data. If there is a problem, and more or new data are needed, you modify your plans as needed and con-tinue with more experiments. With all data in hand, you verify your findings. You draft the re-quested report, and submit it for review. In many cases, the instructors will make comments, and return the draft for revision. (At this stage, it is unlikely that additional time is available for further modifications of the plan and further experiments.) You revise the draft, and submit a final, formal report as described in Chapter 11.

The achievement of any R&D goal by means of the plan discussed above requires the effective use of critical thinking and communication skills. Critical thinking is needed throughout, especially during the planning, execution, and analysis stages. In fact, you cannot successfully follow this paradigm and achieve the goals unless you establish your goals and function in a critical thinking environment. In the modern engineering environment, communication skills are especially needed during the proposal, reporting, and review stages. You will not obtain funding if you cannot clearly express your research agenda. You will not get any of your findings accepted or published if you cannot effectively communicate what you have done.

By virtue of the selection of experiments which you will perform, it is clear that the laboratory course you are embarking on fits perfectly into the paradigm of chemical engineering. You will follow the paradigm of engineering investigation as you plan, execute, analyze, and report.

Part III: Safety—Regulations in the Chemical Engineering Laboratory

Philosophy

Adherence to the rules and regulations described in this section will ensure a safe, productive, and enjoyable engineering laboratory experience. At your university, the department staff and instructors want you to approach the experiments, not with trepidation, but with common sense and a healthy respect for the potential consequence of accidents, especially preventable ones.

Your rule of thumb should always be:

"When in doubt, ask."

No one will ever fault you for asking; on the contrary, you will be admired for considering the safety of yourself and your partners. Remember, safety is an attitude and a habit.

Below are specific guidelines for laboratory safety:

Clothing

- No skirts or shorts; long pants only. Otherwise wear full-length lab coat.

- Tee shirts acceptable, but no sleeveless shirts.

- No sandals, open shoes, or slippers. Sneakers are acceptable, but leather shoes are preferred.

- No loose clothing (e.g. unfastened neckties) or unbound long hair.

- Hard hats must be worn in all high head areas.

Eye Protection

- Safety glasses or goggles are required at all times in the laboratory, except in safe areas far away from apparatus or chemicals (e.g. calculation table).

- Contact lenses are not allowed in the laboratory, even with safety glasses. (Chemicals can get behind a lens and cause serious optical damage.)

Cleanup

- At the conclusion of an experiment, leave the area neat and orderly, paying special attention to the following:

  + Wash off all working surfaces (e.g. tables).

  + Wash floor if needed; remove all excess water; coil hoses.

  + Place lids on all floor drains.

+ Dispose of all loose paper products in trash bins.

+ Wash dirty equipment (e.g. glassware) before returning.

+ Close and secure chemical containers (e.g. bottles, drums).

+ Close and secure all gas cylinders.

+ Remove all weights from scales; secure balance arms.

+ Shut off all utilities (water, electrical, air, steam).

Equipment Repair

• If any lab equipment appears to be faulty, evaluate the situation and make your recommendation to the instructor as to a course of action. Follow the directions of the instructor.

• Students will be allowed to conduct minor on-site repairs. Tools may be borrowed from the stockroom or from the machine shop attendant.

• Upon completion of a repair, obtain the approval of the instructor before resuming experimental operations.

Chemicals

• Exercise caution with hazardous chemicals. Determine specific hazards involved (e.g. read label), and take appropriate protective action. Consult instructor if you are not sure.

• Avoid prolonged exposure to vapors (e.g. organics). Request ample ventilation, if needed. Go outside and get some fresh air. Use fume hood in stockroom when dispensing odorous chemicals.

• Wear safety apron and gloves when handling caustics and acids. Always wear safety glasses under all circumstances.

• Transport individual glass chemical bottles in plastic safety carriers.

• Examine contents of experiment supply carts in stockroom before moving them out. Make sure all chemical containers, especially bottles and equipment, are secure. Exercise caution during transport. Remove all unnecessary materials.

• Do not mouth pipette under any circumstance. Use approved suction bulbs.

Chemical and Waste Disposal

• Solids, Paper, and Glassware

+ Unwanted water soluble, non-toxic solids may be dissolved in water and poured down a floor drain or lab sink. Flush well with water.

+ Bring other unwanted, solid chemicals to stockroom. Inform attendant as to identity and approximate amount of solid.

+ Discard waste paper in trash bins.

+ Place broken glassware in special "Broken Glass" bins.

- Liquids

    + Unwanted water soluble, non-toxic liquids may be poured down a floor drain or lab sink, followed by flushing with large amounts of water.

    + Bring unwanted flammable and other liquids to stockroom. Inform attendant of identity and approximate amount of liquid.

Electrical

- Before using any electrical apparatus or extension cords, perform a quick inspection for any faulty connections or bare wires. Do not use suspect equipment (e.g. bare wires, broken receptacles or plugs); instead, report your suspicions to the instructor.

- Do not allow extension cords to lay on any wet surfaces. Make sure location of any cords are clearly obvious. Obtain approval of instructor before starting experiment.

- Immediately report any sparking or smoking electrical apparatus to instructor. Do not touch such apparatus without approval.

Accidents

- Make sure you know the location of first-aid kits, safety showers, fire blankets, fire extinguishers.

- In case of any accident or fire, do not panic. Inform instructor immediately and follow directions.

- Personal Injury:

    + Minor (e.g. cuts, scraps) — use the first-aid kits.

    + Moderate — escort the injured individual to the campus nurse.

    + Major — call campus police on Emergency Number. Take first-aid action if you are qualified; otherwise, assist in making victim comfortable.

- Fire:

    + If manageable, use fire extinguisher or smother fire with fire blanket. Try to turn off any electrical or chemical supply.

    + If unmanageable, warn others, evacuate lab, call campus police, and activate building fire alarm.

Part IV: Characteristic Problems in Laboratory Performance

Our experience has taught us that our students' success in the laboratory depends upon their adopting a comprehensive research orientation, one that allows them to do the following:

- provide a clear statement of research aims,

- develop appropriate theoretical relationships,

- differentiate between levels of data,

- consider uncertainty limits,

- extrapolate and speculate beyond the confines of the data,

- integrate knowledge, and

- penetrate to the heart of the experiment.

Figure 3.4, you will recall, provided a broad process outline of engineering research. These seven guidelines below incorporate the subtleties of a true research orientation as manifested in the classroom laboratory situation. In their absence, experimentation becomes uncertain; in their presence, experimentation exhibits the characteristics of thoughtful chemical engineering research.

To help you avoid characteristic problems we have discovered in the work of our own students, we turn now to these guidelines. Examples from student reports will be discussed under each guideline.[6]

Guideline 1. Provide a clear statement of your research objective. A clear statement of the objective of an experiment is essential. Here is a superior example from a past student group from the Mass Transfer Experiment described in Figure 3.3:

> The objective of the experiment was to determine the overall efficiency, Murphree stage efficiency, and energy requirements for methanol-water distillation in a ten-tray column.

Unless you can express the aim of your experiment in writing, it is unlikely that you truly understand the experiment. Because we attribute great importance to communication, we believe that you can only understand an experiment if you can express its objective lucidly in writing.

Guideline 2. Develop appropriate theoretical relationships. Often, students incorrectly assume that the purpose of an experiment is simply to verify equations and correlations learned in the earlier classroom courses. For example, in evaluating a Karr column one student group wrote the following:

> The equation used to calculate the number of theoretical stages is
>
> $$N = \frac{\ln \left[(1-1/E)\,(y_2 - m\,X_1)\,/\,(y_1 - m\,X_1) + 1/E \right]}{\ln E}$$

According to . . . Table 3, the results of the analytical method agree closely with the results obtained graphically.

Neither the source nor the theoretical basis for this relationship was identified. The applicability of the theory to the particular experiment at hand was also not established.

While a complete derivation need not be presented, it is important that every presentation of theory include the important assumptions, references, and justifications of the applicability of that theory to the experiment at hand. A schematic diagram which relates the theory to the apparatus should also be presented.

Guideline 3. Differentiate between levels of data. A research orientation requires proper identification of the levels of data. (See Chapter 4 for a further discussion of levels of data.) Many students, however, fall prey to the data dump. By this, we mean that they lump all of the data—ranging from the raw measurements (e.g. titrant volumes) to higher order, derived quantities (e.g. mass transfer coefficient)—together, without indicating their relative importance.

For example, one student group, assigned to study the Karr Extraction Column, presented in the main body of the report the following tables in succession:

Table 1: Titrant Volumes of NaOH and Mole Fraction of Acetic Acid Effluents

Table 2: Titrant Volumes of NaOH and Mole Fraction of Acetic Acid Influents

Table 3: Calculated Karr Column Operating Data

Tables 1 and 2 included raw and low level data—Level 1 and Level 2—which more appropriately belong in an Appendix. Table 3 listed number of stages and column efficiency—Level 3 and Level 4 data—which correctly belong in the main body of the report.

Guideline 4. Consider uncertainty limits. A mature research orientation requires consideration of uncertainty limits. Here is what you must ask yourself before your instructor asks it: Just how solid are my data and the derived results based on them?

It is our experience that students will often not recognize what are and what are not reliable data. Consider a group assigned to evaluate a tubular flow reactor. In order to explain less-than-satisfying results, the group blamed their liquid rotameter calibrations, offering the following recommendation:

> As it was illustrated in the discussion, a major source of error in the experiment was the rotameter calibrations. To improve the accuracy of these calibrations, capacity measuring devices . . . should be installed on the tanks.

This group failed to recognize that measuring the volume of liquid flow through a rotameter with a tared bucket and stop-watch is quite reliable. The group also failed to realize that any sophisticated flow measuring devices would still require calibration in the same manner as the rotameters.

Guideline 5. Extrapolate and speculate beyond the confines of the data. One avenue for the projection of your findings is through recommendations. Below are interesting recommendations from a student group conducting an investigation of yeast growth kinetics:

> The procedure for this experiment should be changed so as to allow a more careful study of the exponential growth period.

This can be achieved by the following method:

1. Start yeast growth as normal.

2. After allowing sufficient time for yeast to be in the exponential growth period, inject batch with fresh feed.

This can be repeated twice during the 10-hour period."

While this suggestion indicates that the group was reflective of their work, they should have gone one step further and indicated how this added step would be beneficial. Nevertheless, it is clear that the group was thinking beyond the confines of the experiment.

Guideline 6. Integrate knowledge. During their discussion of error sources, one group offered the following:

In this experiment, there were two major sources of error: first, the technique of taking samples, and second, their analyses. Two portable heat exchangers equipped with long coils in ice water were used to collect liquid and vapor samples. The coils were not cleaned between samples, so contamination was possible. Partial vaporization of the samples occurred because the sample jar lids were not sealed tightly enough. These errors lead to a decrease in the concentration of the more volatile component, methanol.

This group realized too late that the same relative volatility which allows for the distillation of methanol from water also is active during sample collection and analysis.

As the example suggests, integration of all your knowledge of chemical engineering and other disciplines is important to the critically thinking engineer. Consider the following statement of a group grappling with an experimentally-determined rate constant which was not matching a literature value:

Without taking into effect the 15 °C difference in the literature and experimental reaction temperature, the rate constant should level off . . . .

For a reaction with an activation energy of 30 kcal/mole and a base temperature of 300 K, a 15 degree rise in temperature will increase the rate constant by a factor of almost 11!

Here is a superior example of integration from a group investigating yeast growth kinetics:

There are various reasons why this lag phase occurs. To begin with, the yeast must acclimate itself to its new environment (temperature and pH). Also, certain by-products must be produced by the yeast itself before the cells can multiply. These by-products act as a catalyst for the yeast cell growth. This behavior is similar to that of an autocatalytic reaction.

These statements show an excellent use of references. Important is the last sentences, in which the group recognizes that autocatalytic kinetics represents a potential theoretical approach for analysis of their data. This is a superior example of integrating classroom knowledge into the laboratory.

Guideline 7. Penetrate to the heart of the experiment. Your research orientation is part of your critical thinking orientation. Hence, you must look at the whole experiment and break it down into its fundamental parts. You must determine the essence of what it is you are doing. For example, one group analyzing non-ideal behavior in tubular flow reactors generated the following abstract:

> This experiment studies the effect of the flow rate of the reactants and the size of the tubular reactor on the reaction . . . . The Reynolds number increases from 2985 to 15265 in a 1 inch tubular reactor . . . . The percent conversion of ethyl acetate decreases from 7.2 to 1.4 . . . .

This group failed to dissect the experiment to discover the heart of what they were doing: using an experimental rate constant (which they determined assuming perfect plug flow behavior) as a diagnostic of non-ideal behavior. There is no hint of this in the statements above.

With these seven general guidelines in mind, you will be able to anticipate characteristic problems in your laboratory performance. Yet, a more systematic plan will also be of help to you, so we turn now to a specific troubleshooter's guide.

Part V: The Laboratory Experiment—A Troubleshooter's Guide

The successful completion of an experiment in the chemical engineering laboratory requires proper management. While all students are responsible to the group, the leader must see to it that the efforts of the group are properly directed and all tasks are completed. When planning an experiment, we suggest that you think through these phases:

Planning the Experiment
Executing the Experiment
Preliminary Analysis of Experimental Results
Re-Executing the Experiment
Final Analysis of the Results
Drafting and Reviewing the Report
Revising and Submitting the Report

We will now describe briefly the goals of each phase, as well as offer checklists to ensure that the goals of each stage are being met. You will see that, in performing each of these steps, you are actually taking a critical thinking approach.

Planning the Experiment

You should first become familiar with the goals of the experiment. If you do not understand the goals, the procedure and your observations will have no meaning. Then go over the equipment and its operation, identifying any problem areas and devising solutions. Try to anticipate potential bottlenecks in the experimental procedure. Develop a detailed procedural plan for all the required tasks on a detailed level (e.g., what flow rates to run, how many samples to take, what parameters to monitor). Make sure all needed analytical capabilities are available before starting the experiment.

Probably the single biggest hurdle to overcome in planning is the division of responsibility within the group. This occurs at two occasions: a) during the experimental phase, and b) during the analysis and write-up phase. All tasks (e.g. equipment operation, sample taking, sample analysis) must be assigned before starting the experiment. (See the discussion of collaboration below.)

The group, and especially the leader, must recognize internal strengths and weaknesses. For example, one student might be especially adept at equipment operation, perhaps through a prior summer or co-op job experience. Bear in mind, though, that while division of labor is necessary, each student should fully understand the tasks of each partner.

Checklist on Planning

1. Have you clearly expressed in writing (your lab notebook) the aims of the experiment? (If you cannot do this before the experiment, it is unlikely you will be able to do so afterward.)

2. Have you assessed the equipment to anticipate any problems? Have you established compensatory measures?

3. Have you anticipated any procedural bottlenecks and established expeditious measures?

4. Have all tasks which are necessary for completion of the experiment been assigned? Have the strengths and weaknesses of the group members been taken into consideration?

Executing the Experiment

Execution of the experiment involves more than merely following a procedure. Successful execution of an experiment involves the smooth completion of the tasks as they are determined in your planning stage. Successful execution also requires a thoughtful and critical examination of the observed data as they are collected.

For example, assume that your group has been assigned an extraction experiment. Your preliminary consideration of the objectives and theoretical bases for this process has told you that mechanical agitation is needed in order for the extraction to occur. You suspect that the degree of extraction, as measured by the concentration of solute in the solvent phase, will increase as the agitation becomes more vigorous.

As you proceed with the experiment and analyze your samples, you generate a crude working plot in your notebook of approximate solute concentration in the solvent versus agitation rate. After about three points at low rates which are monotonically rising, the subsequent data points begin to tail off. What should you do?

An approach devoid of critical thinking would be to just forge ahead with more samples. Upon later analyses and write-up, one would blame the equipment, or the old stand-by: experimental error. Because this situation is avoidable, it is unacceptable.

A true research orientation demands that you temporarily stop collecting data and examine carefully those factors which might be affecting your samples: collection procedure, sample work-up, analysis, and so forth. Is some systematic error creeping into your procedure? If you cannot determine a cause, consult the instructor. If you cannot find a cause for error, then continue data collection.

Keep thinking, and look for other factors which you had not planned to consider: e.g., temperature change. Remember to integrate all your chemical engineering background into the problem. The engineer who is thinking critically will not be afraid to consider heat transfer effects during an extraction experiment.

Checklist on Execution

1. Have you developed a working plot of the observed data?

2. Do the trends follow what you might expect? If not, have you considered other effects? Have you compensated for them?

3. Have you estimated all precisions of the measurement and analytical tools used?

4. Have you recorded all pertinent data, your notes, and your thoughts neatly and lucidly in your lab notebook?

5. Have you completed all the experimental tasks established during the planning phase?

Preliminary Analysis of Experimental Results

By the start of this stage, the group leader should have assigned the various tasks of analysis and write-up. You should establish an effective strategy for work-up of the raw data, construction of the appropriate theoretical model, and generation of the higher-level quantities consistent with the experimental objectives.

As you begin to examine your data closely, look for correlations. Remember to establish the levels of data. (See Chapter 4.) Specifying correlations will help you in organizing your efforts. The appropriate theoretical relationships must be developed parallel to critical examination of the data.

When you reach the point where you are comparing your observed correlations with theoretical and/or semi-empirical literature relationships, predictions, and models, critical thinking is truly needed. You should estimate error bars around your data points, and generate uncertainty limits for your final results. (See Chapter 4.)

With some experiments, the results might appear "well behaved" as compared to the model. In this case, the first reaction is often one of relief. Your attitude, though, should not be this: "Wow, my data verify the literature, so I guess I didn't make any mistakes." Rather, the research oriented engineer who has confidence in the data would say: "Wow, I have developed a theoretical model which is consistent with my data within experimental uncertainty limits."

In some cases, you might encounter what seems to be a poor comparison of your experimental results with the theory. Why? There are several possibilities. One recurring cause is the lack of a statistically sufficient amount of data upon which to base a strong correlation. In other words, your results might be just fine, but the trend is lost within the uncertainty limits. Another possible cause is that legitimate errors might have been committed during the experimental procedure. Finally, the theoretical model which you offer might be inadequate to describe fully the experiment. Here, integration of all your chemical engineering background is needed. In case of either of the first two causes, you should correctly procure additional data in the laboratory, if time permits.

Checklist on Preliminary Analysis

1. Have you critically examined your data, establishing levels and estimating uncertainty limits?

2. Have you developed a model to explain your observations?

3. Have you determined all requested results and met the objectives?

4. Have you compared your results with your model, and determined if it is necessary to return to the lab?

Re-Executing the Experiment

If you return to the lab in order to repeat a portion of your experiment or to take additional data, the same guidelines apply as above. Your group should operate even more efficiently than before, and the second session should prove exceptionally fruitful. If time permits, you may even extend the experimental parameters beyond your original plans. This strategy might be suggested by your theoretical developments. Performing additional experiments which might be suggested by theory or extrapolations is an example of a sophisticated research orientation.

Checklist on Re-Executing the Experiment

1. If applicable, have you identified and corrected any procedural difficulties or errors?

2. Have you taken a full set of data sufficient to complete the objectives?

Final Analysis of the Results

Once you have obtained a full set of data and developed a sense of its limitations, you can complete your anaysis and draw final conclusions. An important critical thinking exercise at this point is speculation. Can you extrapolate intelligently with your results? Another important task here is to develop realistic recommendations regarding the experiment just completed, especially regarding procedural matters.

Checklist on Final Analysis

1. Have all objectives been met? Have all final results and conclusions been generated?

2. Have you critically evaluated your results? What degree of certainty can you affix to them?

3. Are your speculations/extrapolations reasonable?

4. Have you developed realistic recommendations?

Drafting and Reviewing the Report

An appropriate reporting structure is as important a part of the laboratory experience as the experiment itself. If you cannot successfully communicate your findings to others, then your findings are worthless and your efforts will have been wasted. Successful articulation is your primary means of demonstrating your ability to think critically.

As part of the planning process, tasks must be assigned for drafting the report. Following the guidelines in Chapter 9, create your draft. Do as good a job as possible, since instructors frequently adopt a policy of not offering revisions for initial reports receiving a grade lower than "C." Make sure each group member thoroughly reads through the entire document. This will ensure continuity and internal consistency.

Submit a draft to a peer group and to your instructor for preliminary grading and comments. The draft will then be returned for revision. This iteration process, we believe, is important to the laboratory since experimental knowledge is itself iterative—you are never finished learning.

Checklist on Drafting and Reviewing the Report

1. Have you followed all the guidelines for the particular reporting structure required?

2. Have all group members completely reviewed the draft and approved it?

3. Have you extended an effort worthy of a final draft?

Revising and Submitting the Report

This stage represents your last opportunity to generate a top-notch product. Carefully consider the instructor's comments and implement corrections. Again, review the guidelines offered in Chapter 9 and the laboratory reporting structures in Chapter 11.

Checklist on Revising and Submitting the Report

1. Have you seriously considered the comments of your instructor and your peers and incorporated them into your final document?

2. Have you generated a professional final draft?

Part VI: Collaborative Work

An important component of the critical thinking environment, and a key to your successful completion of the laboratory, is a healthy collaboration with the members of your group. In the past decade, a great deal of research has been done on ways that professionals collaborate.

For example, Lisa S. Ede and Andrea A. Lunsford conducted a study of collaborative writing among members of the American Institute of Chemists, the American Consulting Engineers Council, the American Psychological Association, the International City Management Association, the Professional Services Management Association, the Society for Technical Communication, and the Modern Language Association.[7] The researchers discovered that a large majority of those studied write collaboratively and that 59% found collaborative writing either "very productive" or "productive." In addition, almost 98% of those surveyed believed that effective writing was "very important" or "important" to their jobs. What are the conditions for successful collaboration? The researchers identified effective group dynamics, shared goals, and efficiency as fundamental characteristics of effective collaboration.

David W. Johnson and his co-authors offer a four-step method for effective collaborative work.[8] We list these four steps below, supplemented by our own knowledge of what helps our students work well together:

Step 1. *Forming: Assignment of Responsibilities.* First, each group must have a leader to oversee the experiment as a whole and to ensure that all tasks are carried out. (The role of group leader, we believe, should rotate over the course of the semester so all experience the responsibilities of management.) The group, guided by its leader, must assess the strengths and preferences of its members in order to assign responsibilities. This brief step, lasting no more than thirty minutes, will strengthen the group effort and ensure that individual efforts will be equivalent. A tried-and-true method of task assignment follows:

Equipment Operation (one student)

Sample Collection (one student)

Sample Analysis   (one student)

Step 2. *Functioning: Preview of Goals.* In a one-hour meeting, the group members should preview the experiment to determine efficient and effective methodologies, to review each member's responsibilities, to familiarize all members with all facets of the experimental procedure, and to anticipate problems and their resolution. Johnson suggests that the group work during this meeting to establish a time line for meeting various objectives.

Step 3. *Formulating: Assessment of Experiment.* The group's assessment of its efforts should be on-going—assessment should be formative, not summative. For example, while the experiment is being conducted, the group must determine when they have gathered sufficient data or whether or not to re-execute their experiment. After all data has been gathered, the group should meet for approximately two hours to seek accuracy of results, elaboration of results, and creative ideas. During this assessment phase, the group also must identify experimental weaknesses and plan effective solutions.

Step 4. *Fermenting: Final Assessment.* This stage includes completing calculations and writing the report; it could take 10 to 20 hours of work. In many universities this stage includes an expert review by the instructor before the final report is submitted. We elaborate on this crucial aspect of collaborative work below.

Part VII: Assessment in the Chemical Engineering Laboratory

The evaluation of your work is a crucial part of your success in chemical engineering. It is our experience, however, that students do not set clear goals of acceptability for themselves; rather they rely far too heavily on their instructor's evaluation; hence, they often fail to become independent thinkers of the kind specified in Chapter 1. Your reports will be far better if you adopt another system of evaluation.

We suggest that you explore evaluation as shown in Figure 3.5 below:

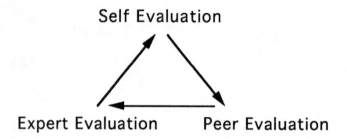

Figure 3.5 Three Kinds of Assessment in the Chemical Engineering Laboratory

Self assessment is the process by which you critically evaluate your own work.

Peer assessment is the process by which others similar to you in experience and expertise evaluate your work.

Expert assessment is the process by which a specialist in your field evaluates your work.

What should you look for in assessing your work and the work of others? In the chemical engineering course, the key to assessment rests in fulfilling well-specified, precise goals for each experiment and reporting your fulfillment of those goals in writing.

Evaluation sheets such as the one shown below in Figure 3.6 are very useful in the chemical engineering laboratory. This sheet specifies the criteria for the experiment described in Figure 3.3 above.

Figure 3.6 A Sample Evaluation Sheet for a Mass Transfer Experiment

MASS TRANSFER EXPERIMENT # 4
CONTINUOUS DISTILLATION OF A BINARY MIXTURE

Below are listed selected goals of the distillation experiment. After reading the group's report on the experiment, please circle the response indicating your judgment as to whether the group has exhibited

(more)

4 - Superior ability
3 - Competent ability
2 - Somewhat competent ability
1 - Lack of competent ability

+ OBJECTIVE

Determine the separation efficiency of a ten-tray distillation column processing a methanol/water mixture at two different reflux ratios.

4   3   2   1

+ BACKGROUND

Briefly explain the meaning and use of the McCabe-Thiele and Ponchon-Savarit methods to obtain the number of theoretical stages; distinguish between ideal and actual stages; analyze relationship between Murphree tray efficiency and column efficiency.

4   3   2   1

+ PROCESS

For one finite and the infinite reflux ratios, obtain appropriate vapor and liquid composition samples.

4   3   2   1

+ DISCUSSION

Address the following items: a) representative temperature/tray profiles; b) the actual separation (i.e. composition of top and bottoms products as compared to feed); c) graphical determination of ideal stages; d) tray efficiencies; e) McCabe-Thiele method vs. Ponchon-Savarit methods; discuss why they yield similar results here.

4   3   2   1

+ CONCLUSIONS

Provide statement regarding the efficiency of the column in separating the mixture; provide any useful recommendations.

4   3   2   1

Two sheets should accompany each preliminary evaluation of your work: one completed by a peer group evaluating your experiment, the other by your instructor. If these sheets are filled out before the final report is submitted, your chance of success is greatly enhanced. By this method, you will be able to pinpoint your problems and address them with precision.

You should also note how your scores improve on successive drafts, and you should be able to establish the reasons for that improvement. All assessment sheets should be kept, along with copies of your reports, as a record of the kind and amount of review your reports received and how these reports improved. These sheets may then be retained by the instructor at the end of the course and used by your department for review by its industrial advisory board and by the Accreditation Board for Engineering and Technology (ABET).

Assessment requires the kind of research orientation we discussed in Chapter 1. By thinking critically about your reports and experiments, you will produce stronger, more powerful documents. By becoming a better evaluator of your work and the work of others, you will become a better chemical engineer.

Conclusion to Chapter 3 : What is the Laboratory (Really) About?

With all the advice above in mind, it is worthwhile for you to realize what is really behind your course. The purpose of a chemical engineering laboratory course goes beyond your verifying the theoretical and semi-empirical relationships that you learned in your junior– and senior–year classroom courses.

You are engaged in your laboratory course in order to learn to think critically about your field and to communicate your work in that field effectively. If you therefore establish a critical thinking environment, you will be ready for industry, graduate school, or whatever profession you choose because you will have learned to think and to communicate.

You should approach an experiment as if you are the first to have ever operated it. Cultivate the sense of curiosity that brought you into chemical engineering. After careful examination, you will see that there is nothing magical about the apparatus. Valves are connected to piping; switches are connected to wires. Follow the guidelines outlined above for planning the experiment. After the experiment is run, critically examine your data, establishing correlations between variables, acknowledging the limits of certainty. Develop, then, a theoretical model consistent with your observations. Finally, make conclusions and recommendations in a well–organized reporting structure.

In following these steps, you will demonstrate not only your understanding of the classical paradigms of chemical engineering, but also your ability to think critically about your field.

Notes and References

1. David A. Hounshell and John Kenly Smith, Jr., *Science and Corporate Strategy: DuPont R & D, 1902–1980*, New York: Cambridge University Press, 1988.

2. Quoted in Hounshell and Smith, p. 3.

3. Alan I. Marcus and Howard P. Segal, *Technology in America: A Brief History*, New York: Harcourt, 1989, p. 146.

4. Thomas Kuhn, *The Structure of Scientific Revolutions*, Chicago: University of Chicago Press, 1962.

5. There has been much scholarly discussion on the research process in recent years. For further discussion, see the following: Barry Barnes and David Edge, eds., *Science in Context: Reading in the Sociology of Science*, Cambridge: MIT Press, 1982; Henry Petroski, *To Engineer is Human: The Role of Failure in Successful Design*, New York: St. Martin's Press, 1985; Bruno Latour and Steve

Woolgar, *Laboratory Life: The Construction of Scientific Facts*, Princeton: Princeton University Press, 1986; Bruno Latour, *Science in Action*, Cambridge: Harvard University Press, 1987; Ruth Bleir. ed., *Feminist Approaches to Science*, New York: Pergamon Press, 1988.

6. An approach similar to ours is taken by Wallace B. Whiting in "Errors: A Rich Source of Problems and Examples," *Chemical Engineering Education,* (Summer 1991), pp. 140–144.

7. Lisa S. Ede and Andrea S. Lunsford, "Collaborative Learning: Lessons from the World of Work," *WPA: Writing Program Administration*, (Spring 1986), pp. 17–26.

8. David W. Johnson et al., *Circles of Learning: Cooperation in the Classroom*, Association for Supervision and Curriculum Development, 1984.

Six Assignments for Juniors and Seniors Enrolled in Chemical Engineering Laboratory Courses

Below are six assignments that will help you think critically about your laboratory courses.

Assignment 1

Refer to Figure 3.1, and place the experiments in your lab within the figure. Does Figure 3.1 accommodate your lab? Why or why not? What does the distribution of experiments suggest about the kinds of chemical engineering experiences you will have in your lab?

Assignment 2

Refer to Part III of this chapter, and make a safety checklist of possible hazards of each experiment before you perform it. Be sure to record the location of all first-aid equipment in the laboratory.

Assignment 3

Review the seven guidelines provided in Part IV of this chapter, and design a checklist of characteristic problems in laboratory performance that you may use in all of your experiments in the laboratory course. Have your instructor check the list and comment on its strengths and weaknesses.

Assignment 4

After you perform your first laboratory experiment, design a troubleshooter's checklist based in the information found in Part V of this chapter. Be sure, however, to tailor the checklist to your particular experiment. Show this checklist to your instructor and find out if you have correctly anticipated all of the most problematic aspects of the experiment.

Assignment 5

Think about the best and worst experiences you have had in collaborative work. Do the best experiences follow the information given in Part VI of this chapter?

Assignment 6

As you begin each experiment, draft an assessment sheet similar to that shown in Figure 3.6. Show this sheet to your instructor for revision and use the revised sheet for self, peer, and instructor evaluation. Note how these kinds of review are similar and different. Did the assessment sheet (1) help you identify weaknesses in your work and (2) help your group write a better report?

Experiment vs. Theory: Which Do You Believe?

As you compare the predictions of your theoretical model with your experimental data, bear in mind that this step is a powerful one. If your model results compare well with the data, you are well on your way to achieving an understanding of the process under study. This situation can be represented as in the diagram below:

Experiment ⟶ Model ⟶ Truth

This relationship, though, is somewhat idealized. More times than not, the real situation can be represented as this:

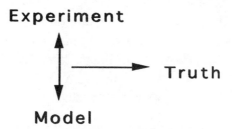

Not only can the experimental data suggest an appropriate model, but the reverse often occurs. If you have confidence in your model— especially if it already does a good job explaining other data—then a lack of agreement between the model and your current data may suggest that the data are in error. The following true example occurred recently in Dr. Barat's own laboratory.

> We were attempting to measure relative concentrations of OH radical in a laminar flat flame using laser induced fluorescence (LIF). Data from the methane (CH_4)/air flame were consistent with those published in the literature. The big question? How would those LIF data change when methyl chloride (CH_3Cl) was introduced into the feed mixture?

(more)

Preliminary results showed a decrease in the LIF signal. We expected this. We were elated, and we began work on writing a major journal article.

Additional experiments in our lab were performed in which concentrations of stable species such as carbon monoxide (CO) and carbon dioxide (CO_2) were measured in these flames by gas chromatography. These data also exhibited trends which went the way we expected. All seemed well with the world.

Computer calculations began with a theoretical model to simulate the experiments. The predicted concentrations of the stable species such as CO and CO_2 compared well with the data. However, the model predicted a factor of 10 reduction on OH concentration when CH_3Cl was fed into the CH_4/air flame. The LIF data showed less than a 25% drop. We were stunned! What was going on? Which should we believe—the model or the data?

Realizing that optical work can be especially difficult, we took a closer look at how we were performing the LIF experiment. After careful scrutiny, in conjunction with additional reading of the literature, we discovered that the LIF experiment was not being done properly. Immediate changes were made. And we would have to hold off for a while on that journal article.

The real issue here is not that the LIF experiment was wrong. The important lesson is that a model—if reasonable and supported by established data—can say something about the validity of your data.

The point: the interaction between data and model is dynamic. It goes both ways. It is a very important element of the critical thinking environment. It is characterized by shifts in confidence as you recognize the limits of certainty in your data (Chapter 4) and the difficulty in modeling complex phenomenon. Indeed, how you handle this dynamic relationship is also an issue in laboratory ethics (Chapter 6).

4 The Uses of Argument in Chemical Engineering

"The world does not speak. Only we do."
Richard Rorty, *Contingency, Irony, and Solidarity*

"Doubt is not a pleasant state but certainty is a ridiculous one."
Voltaire

Preview

In Chapter 4, you'll have a chance to think about

- the meaning of scientific objectivity and its limits

- a method for presenting your arguments

- methods of examining your data and analyzing your results

"And for that I shall tell you that in ancient times a debate hath risen—and it remains yet unresolved—whether the happiness of man in this world doth consist more in contemplation or action."

Isaac Walton, *The Compleat Angler*

The Tendency Toward Objectivity

Engineers strive toward truth and objectivity. Working within the limits of physical law, you are bound ethically to be as objective as possible.

Nevertheless, your performance will be influenced by a degree of subjectivity based on your own background, previous training, and interests. In fact, research in learning tells us that how we assimilate new information will depend on the kinds of information and knowledge we already have. In addition, the very empirical methods we use to observe natural and physical phenomena are limited in precision. To top off our dilemma, the language by which we describe our findings is itself often ambiguous. As the philosopher Richard Rorty puts it," The world is out there, but descriptions of the world are not."[1]

How are we to understand these limits of self, measurement, and language?

Let us begin with an example, typical of what we have seen over the last few years in our own laboratory, of how student performance can be limited by subjective assumptions:

> A group studied the hydrolysis of acetic anhydride in a non-adiabatic, non-isothermal batch reactor. The very simple procedure involves obtaining a temperature-time trace. (The elegant theoretical analysis of the data generates kinetic and thermochemical parameters.) When the students came up with kinetic parameters which differed significantly from accepted literature values for this reaction, their immediate comment to us was: "The procedure was so simple; we've checked our analysis, and we still can't get the results to match the literature. We can't figure out what went wrong."

> We then examined the reactor thermocouple and found it to be in a very murky sheathed oil bath. We suggested to them that their temperature/time data was really a convolution of the true kinetic trace and the response transfer function of the sheathed thermocouple. Their response: "Wait a minute, that's a topic from process control class, and this is not the process control lab."

The students' unwillingness to bring all of their chemical engineering knowledge and general experience to the experiment prevented them from performing well. By compartmentalizing, rather than integrating, knowledge, they limited their ability to analyze the hydrolysis of acetic anhydride in a non-adiabatic, non-isothermal batch reactor.

The Limits of Knowledge

Physics has shown that there are limits to how certainly we can know something. A classic example, of course, is the Heisenberg uncertainty principle set forth in 1927. Heisenberg had discovered that the behavior of subatomic particles is more a matter of statistical probability than of determinable cause and effect. Because it is impossible to measure accurately certain pairs of quantities at the same time, we can infer that no events or objects can be described with zero tolerance. Much that had seemed certain about the physical universe only a half a century before now seemed problematic.

Consider also the following insightful example of the limits of knowledge described by J. Bronowski in *The Ascent of Man:*[2]

> A man is examined in a step-wise fashion by shining various ranges of the electromagnetic spectrum at him, from radio waves up through X-rays. What is observed is that each

wavelength range will offer some pieces of information about the man, but not the entire description. We are frustrated in our attempts to gain absolute knowledge of the man.

Bronowski explains that, in our attempts to learn, we confront the "paradox of knowledge." No matter how precise our instruments, we are ultimately limited by a tolerance. In effect, there is no absolute knowledge.

Consider a laboratory example. You are measuring a length of pipe with a tape measure. The smallest divisions are 1/16 inch. A moment of reflection will convince you that your ability to know exactly the pipe length is limited by the smallest division of the scale. The best you can do is to make many repeated measurements and treat them statistically. Karl Gauss, in the 1800's, told us that the mean of your measurements is the most probable estimate of the true length, which lies somewhere within his classic curve.

How, therefore, do we achieve (or approach) certainty? The answer lies with engineering and scientific argument.

A Strategy for Argument: Toulmin's Logic

You must realize that you are not alone in this world of blind spots and uncertainties. Everyone else is similarly limited. Indeed, it is the recognition of these biases—not the wholesale abandonment of objectivity—that is key to understanding the outcome of the scientific method. Only when we recognize our blind spots can we strive to overcome them and move toward truth. Or, put another way, the lack of absolute knowledge does not mean that we cannot be sure about anything.

Your success, therefore, will very strongly depend on your ability to argue your point of view convincingly. Since all chemical engineers labor with the same paradigm, and use the same tools, consensus and ensuing acceptance is possible.

A powerful strategy for effective engineering argument may be found in an adaptation of the analytic system proposed by physicist Stephen Toulmin.[3] A version of his method is pictured schematically in Figure 4.1:

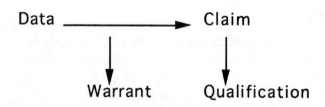

Figure 4.1 Toulmin's Logic

First, you gather all your *data* to support the *claim*, which is your argument. The *warrant*, either implicit or explicit, relates how appropriate the data is to the claim. Finally, the *qualification* supplies any reservations you might have about the claim.

Let's consider the following example from the laboratory:

A group of our students evaluated a shell-and-tube heat exchanger. After collecting various data and applying an appropriate heat balance theory, they generated a correlation of heat transfer coefficient as a function of Reynold's number (the *data*). They observed that the coefficients were numerically lower than those predicted with a literature correlation. They *claimed*, or argued, that this meant that the tube walls were dirty. They felt that this claim was *warranted* because their coefficients were all consistently lower than the literature, but followed the same trend. Also, modification of their theoretical model to account for the added resistance to heat transfer by dirty scale would provide a closer fit to the data. The students *qualified* their claim by admitting that they were unable to open up the exchanger to visually verify pipe scale.

The experiments which you will perform in your laboratory course are vehicles for investigating the chemical engineering paradigm presented in Chapter 3. Simply running the apparatus, collecting data, comparing to theory, and writing the report are not what your course is all about. You are investigators, examining an issue in a critical thinking environment, with your efforts culminating in a reporting structure where you will, in effect, present your argument.

Effective argument requires that you organize and critique your data. These two topics are considered in the next two sections.

Levels of Data

There are four levels of data (or quantities) handled in laboratory reporting. These are described below.

Level 1. These data are the lowest order quantities. They are raw data obtained directly during the experiment which, by themselves, are essentially meaningless. They are usually not presented in the main body of any report, written or oral, but may be presented in an appendix. Examples include titrant volumes, rotameter settings, and voltages.

Level 2. These quantities can stand alone and have meaning. They are generally simple quantities derived directly from Level 1 data, but they can also be raw data obtained from direct reading instruments. These data can be placed in the main body of a report; each is usually presented as the independent variable in a plot or correlation versus a Level 3 or Level 4 quantity as the dependent variable. Examples include mole fractions, flow rates, and temperatures.

Level 3. These quantities are derived from lower-level data, usually level 2. Such values not only can stand alone, but often have implications beyond their face value. Level 3 values are definitely main body material. Examples include conversions, Reynolds numbers, and column efficiencies.

Level 4. These values are of the highest order, derived from Level 2 and Level 3 values. These main body quantities are generally final results upon which conclusions and recommendations are made. Level 4 values are often correlated with Level 3 quantities. Examples include rate constants, numbers of stages, and transfer coefficients.

The above delineation is meant as a guideline. There are arguable overlaps between levels. What is important is that you, as the experimenter, must identify what is crucial to the argument and then highlight those points in the main body of your report. Supportive data and quantities which are really not directly needed should be located elsewhere in an appendix.

Effective discrimination of your data can only strengthen your argument. There are few things that will alienate your audience faster than a data dump wherein all data are lumped together, an act indicating that you really do not know what is important and what is supportive—and therefore perhaps do not know what you are doing.

The Use of Uncertainty Bars

As part of the analysis of laboratory data, it is important to estimate how "good" the final calculated results are. This analysis, to return to Toulmin's scheme, represents a quantitative *qualification* of your claim.

To begin, remember that there is a degree of uncertainty—a contingency of precision—associated with each measurement. These individual, statistical errors propagate through to the final result. If you think about the levels of data discussion above, you will realize that the error associated with Level 1 and Level 2 data will accumulate through Level 3 values to produce the overall uncertainty in Level 4 values.

Consider the following relationship:

$$z = f(x,y,w) \qquad\qquad (1)$$

The variables x, y, and w are experimental quantities, generally Level 1 and Level 2 data. The quantity z is a Level 3 or Level 4 value. The differential dz is given by:

$$dz = (df/dx)\,dx + (df/dy)\,dy + (df/dw)\,dw \qquad\qquad (2)$$

The factors df/dx, df/dy, and df/dw are partial derivatives obtained analytically from Eq.1. The quantities dx, dy, and dw represent the experimental precision of the measurements, which must be determined or estimated. Unless it is known otherwise, it is assumed that these errors are random (i.e., there are no systematic errors present).

The maximum probable error (MPE) in z is then given by:

$$MPE(z) = dz/z \qquad\qquad (3)$$

An error analysis is appropriate for any experimental effort and its report. The MPE should be determined for your final Level 3 and Level 4 results, as appropriate. Graphically, the MPE should be shown as quantitative +/- uncertainty bars. If listed in a table, the MPE should be given as +/- values after the reported result. Remember that the MPE as calculated from Eq. 3 is a fractional quantity. It might be more appropriate to report dz with the correct units instead.

If you had enough time to take a statistically large number of measurements of a single quantity (e.g. 20 readings of a thermometer for a single temperature reading x), the mean value would represent the best estimate of the "true" value. The uncertainty dx in the reading would be given by the standard deviation of the measurements. However, time constraints make such repeated measurements impossible. You must therefore make an estimate of dx.

A reasonable estimate of the uncertainty (dx, dy, etc.) or precision in a measurement is one-half of the smallest division you can read on the measurement scale. For example, for a thermometer with smallest divisions of 1 degree, the precision can be estimated to be 0.5 degrees. Good judgment must be used in estimating the precision. You might feel that the best you could reproduce a measurement is within one full division of the smallest scale.

In summary, if you are to have confidence in your arguments in the laboratory, you must be prepared to account statistically for the *claim* that you make.[4]

Going Out on a Limb

An important aspect of the critical thinking environment in the chemical engineering laboratory is the ability to speculate. Based on your data, results, warrants, and qualifications, you should be willing to think beyond the confines of the current experiment, and speculate on directions for further research. Your willingness to do this will indicate how strongly you feel about your argument.

Obviously, the further you walk out on that limb, the odds on your falling increase. Therefore, it is prudent to extrapolate to only a small degree. Yet, this extrapolation is appropriate in that, regardless of what the popular voice may say, the body of scientific knowledge grows in small degrees, not in huge leaps. Your laboratory experience—with all its contingencies— is not an end in itself, but a beginning.

Notes and References

1. Richard Rorty, *Contingency, Irony, and Solidarity*, New York: Cambridge University Press, 1989, p. 5. For more on the limits of objectivity, see the following: Charles Coulston Gillispie, *The Edge of Objectivity: An Essay in the History of Scientific Ideas*, Princeton: Princeton University Press, 1960; Evelyn Fox Keller, *Reflection on Gender and Science*, New Haven: Yale University Press, 1984; James Gleick, *Chaos: Making a New Science*, New York: Viking, 1987.

2. Jacob Bronowski, *The Ascent of Man*, Boston: Little, Brown, 1973.

3. Stephen Toulmin, *The Uses of Argument*, New York: Cambridge University Press, 1964.

4. For more on analysis of data, see P. R. Bevington, *Data Reduction and Error Analysis for the Physical Sciences*, New York: McGraw-Hill, 1969.

An Assignment for Juniors and Seniors: Analyzing an Argument in Chemical Engineering

With your instructor's help, select two articles in your field. The first should be a technical article for specialists in your field, and the second should be an article for a broader audience. For example, you might select a technical article from *Industrial and Engineering Chemistry* and an informative article from *Chemical and Engineering News*. After reading both articles, answer the following questions with your lab group and your lab instructor.

What claim is made in the technical article? How is that claim established? Is the relationship between the data and the claim—the warrant—implicit or explicit? Why is this relationship stated or assumed? Are there any qualifications to the claim?

What claim is made in the informational article? How is that claim established? Is the relationship between the data and the claim—the warrant—implicit or explicit? Why is this relationship stated or assumed? Are there any qualifications to the claim?

What can you infer about the way that engineering argument takes place in a technical forum? How important are data? Is there discussion of the limits of data? Does the article go out on a limb? How far does it go?

What can you infer about the way that engineering argument takes place in a less technical forum? Are data important here, and is there discussion of their limits? Or does the discussion center more on the impact of a technology? Does the author assume technical expertise? If so, how can you tell? If not, what strategies are used to carry forth the core information?

Sidebar

Uncertainty Bars Really Make a Difference

An estimation of the degree of certainty of your experimental results can do more than simply improving the relative "fit" of your model to the data. The use of uncertainty bars can offer a kind of insight that will enable you to make more subtle and more accurate conclusions.

Consider Figures 1 and 2 below. Both represent experimental results from two different student groups who performed a tubular flow reactor experiment. Isothermal conversion data, as a function of flow rate for two reactors of different length/diameter (L/D) ratios, were analyzed with a plug flow reactor model. Apparent reaction rate constants were obtained and plotted as a function of Reynolds number.

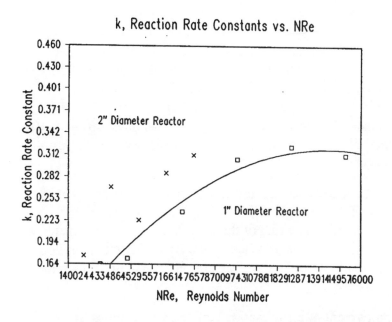

Figure 1. Apparent Rate Constant vs. Reynolds Number (without uncertainty bars)

(more)

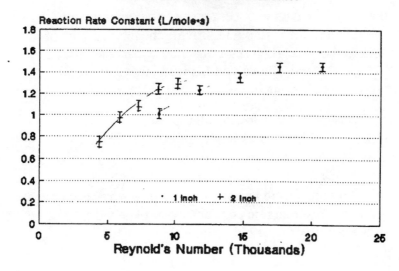

Figure 2. Apparent Rate Constant vs. Reynolds Number (with uncertainty bars)

In Figure 1, while there is clearly a trend, it is difficult to see any real difference in the dependence of k on Reynolds Number between the two reactors. If there is any dependence, it is washed out by what appears to be significant scatter. (In addition to a lack of uncertainty bars, Figure 1 has other problems, most notably an incomprehensible x-axis and no units for k on the y-axis).

In Figure 2, however, a real difference can be inferred. That is, the difference between the curves from the two reactors is clearly more than the width of the uncertainty bars. This difference suggests that a dependence on the Reynolds Number alone is not enough to describe the apparent rate constant. In fact, the L/D ratio is also important.

The employment of error bars will enable you to have greater confidence in your experimental results. As well, error bars will allow you to make the kind of subtle judgments required in both academic and industrial settings.

5   Conducting the Literature Search

Anne M. Buck
University Librarian

"They who contemplate a few things have no difficulty in deciding."

Aristotle

Preview

In Chapter 5, you'll have a chance to think about

- various strategies that will help you review research within and beyond chemical engineering

- planning specialized laboratory research and identifying various sources of information retrieval

- strategies for keeping up-to-date on technological developments in related fields

"I undertake to acquaint the reader with many things that are not usually known to every Angler."

Isaac Walton, *The Compleat Angler*

Part I: Conducting the Literature Search

Introduction to Part I: The Importance of Others' Research

The work of a chemical engineer requires the ability to plan, to solve problems, to develop processes, and to design useful products. Whether in a university unit operations laboratory or in an industrial setting, successfully meeting a technical challenge demands knowledge of the underlying scientific and mathematical principles. (As Aristotle suggests, it is easy to make up your mind if you only contemplate a few things; the contemplation of many things is far more difficult.)

Success, however, also demands awareness of the research of others. The engineer who is able to acquire and apply information to the decision-making process can avoid wasting time on guesswork, reduce the chance of duplicating the work of others, and improve the quality and effectiveness of proposed solutions. Hence, the ability to perform a literature search becomes part of the critical thinking framework established in Chapter 1.

There is no lack of information available in chemical engineering and related fields. This chapter will describe different information sources, their focus, and how they may be accessed. At first, the wide variety of options may seem overwhelming; the body of chemical-related information is enormous and growing exponentially. However, as you discover the value of these sources and apply their contents, you will become familiar with the tools of your discipline.[1]

Library Assistance

Your university library collects many of the information resources that you will require as a chemical engineering student. Your reference librarian will guide you to these resources and will assist you in obtaining materials that may be located outside of your library. In an era of rapidly expanding technical literature and shrinking book and journal budgets, your librarian's knowledge of alternative resources and search strategies will prove invaluable.

Guides to Finding Information

There are a number of reference guides to the process of researching chemical information. These guides are especially useful for their discussion of searching techniques and their coverage of classic sources. These guides, however, quickly become dated as new sources arise and new technologies emerge to store, retrieve, and transmit technical information. The reference librarian can help select an appropriate title and can verify names and addresses of specific information sources that are listed.

Preserving Background Information

Engineers and researchers have discovered that the personal computer (PC) is an ideal tool for building and manipulating files of useful references from the technical literature. Personal indexes on a PC help prevent the loss of hard-to-find information such as physical property data and sources; and, PC files may be searched and sorted in a variety of ways. Commercial software is available for managing the following aspects of your research:

- Information search strategies

- Search results

- Bibliographic citations

- Physical properties data and sources

- Research notes

- Personal papers, reports, and correspondence

If your PC is networked, you can share the contents of information files easily with colleagues. No matter how you decide to use them, the creation of personal information databases is a career-long activity. Indeed, the use of these databases is characteristic of the research orientation described in Chapter 3.

Planning Your Search Strategy

As a chemical engineer, planning your search strategy might be the most critical element in discovering necessary information. Because such an extraordinary amount of chemical-related data exists, an effective strategy will conserve effort, time, and money. R. G. Maizell suggests that you ask yourself these questions before you begin your search:[2]

1. Are my goals and objectives clearly defined?

2. What information is already on hand?

3. How soon do I need this information?

4. How important is information to my project?

5. What time period do I need to cover?

6. Do I need to seek information from international sources?

7. Can my search be limited to a specific type of documentation (i.e., patents)?

8. What specific aspects of the field am I interested in?

9. What sources are likely to provide the most relevant information?

10. How available are these sources?

11. Do I have enough time to contact individuals and receive their responses?

12. Have I checked sources that provide quick answers? For example, can I find the answers I need in a reference book instead of *Chemical Abstracts*?

13. What subject key words are appropriate to my area of inquiry?

14. Am I keeping a systematic record of the elements of my search?

15. Do I know when to stop?

Answering such questions before you begin your search will enable you to develop a process of conducting a literature search that is efficient and fruitful.

Manual Versus Electronic Searching

Until machine readable databases became available, all literature searches were done manually; that is, appropriate texts were identified and reviewed individually. In the 1960s, however, technical publishers began switching to photocomposition and other computer-aided production techniques that placed computer searchable information on magnetic tape. Since then, rapid advances in computing systems, software development, and telecommunications have made electronic searching an important way to obtain information.

However, searching online is not the only way to retrieve useful results. Often, an online search might not be the best approach to the literature. Manual searching, you should know, continues to offer important advantages:

- *Scope*: Some important sources are available only in print.

- *Depth*: Printed versions are complete: online indexes usually cover only the period from 1970 to the present.

- *Cost*: Once acquired, there are no further charges for accessing print sources; online information services are expensive and libraries often charge these costs directly to users.

- *Ease of Use*: It is easy to use a printed index and regular use is not required to maintain search skills.

- *Browsability*: Many important insights have resulted from the chance discovery of a relevant citation outside the scope of the search but on a nearby page.

Sources of Information

Chemical literature appears in a wide variety of forms and formats. Knowing the focus and contents of basic sources saves search time and will lead to better results. In the following sections are discussions of these forms and formats.

Technical Journals

Journals, often referred to as the "primary literature," remain the chief vehicle for the communication of technical and scientific findings. Their value is based upon broad availability, regular publication, peer review, and ongoing collegial dialogue through letters. As research output increases, the number of journals is proliferating, and they are becoming increasingly expensive and more highly focused.

Journals published by societies, research organizations, and universities emphasize original research; examples are *Chemical Engineering Progress* and the *AIChE Journal*, both published by the American Institute of Chemical Engineers. A wide variety of journals are also published commercially, such as the McGraw-Hill publications *Chemical Engineering* and *Chemical Week*.

And journals such as *High Technology Business* are published by trade or manufacturing associations.

Professional societies, special interest groups, and technical publishers also produce extremely subject-specific journals, often as "letters." Such journals are valuable because they not only contain descriptions of new findings but also provide a forum for lively dialogue among colleagues.

Conference Proceedings

Proceedings are a valuable source of information that is more up-to-date than the journal literature. Although these papers might not be refereed by peers, they often contain extensive hard data and are covered by indexes. Libraries collect proceedings of conferences, symposia, workshops, and other compilations of published meeting papers.

Technical Reports

Technical reports contain advanced information on the progress and results of research and development. They may be issued by any type of organization or agency and are valuable because they appear ahead of journal articles. Often this information is proprietary, but even when it is not, technical reports are difficult to identify and locate outside of one's own network of colleagues.

Covering most government sponsored research, The National Technical Information Service (NTIS) publishes the *Government Report Announcements and Index* in print and provides *NTIS Online*. All reports of university or industry research, however, might not be included. Your reference librarian has the names and addresses of additional sources of government technical information.

Indexes and Abstracts

Cumulative indexes such as *Chemical Abstracts (CA)* and *Engineering Index (EI)* bring together the output of a broad collection of technical journals for a particular period of time. An indexer assigns standardized descriptive subject headings to each entry so that all entries on a given topic will appear together. Libraries usually list indexes and abstracts along with the periodicals because they are handled much like journals, are usually located near them, and provide access to their contents. The major indexes include abstracts that provide enough information to decide whether the full article is needed; indexes also make keyword searching in the electric versions very powerful.

It is relatively easy to learn to use print indexes. Publishers issue guides for researchers that provide notes on organization, terminology, journals covered, and so forth. For example, the *Chemical Abstracts Index Guide* appears annually and includes an Introduction, Alphabetic Sequence, Volume Indexes, Selection of General Subject Headings, Chemical Index Substance Names, and Hierarchies of General Subject Headings.

Chemical Abstracts, covering all fields of chemistry and chemical technology, includes the Weekly Issue Indexes and six semiannual indexes (Chemical Substances, General Subjects, Formulas, General Substances, Authors, and Patents). In addition, *Chemical Abstracts* publishes

Ring Systems Handbook, Registry Number Handbook, and the *CA Service Source Index (CASSI)* besides the *Index Guide*. *CA* is also available online.

Engineering Index provides extensive coverage of worldwide literature in all engineering disciplines with full abstracts of journal literature, technical reports, conference proceedings, and monographs. It is available online and on CD-ROM as *Compendex*Plus*™; a subset, *EiChemDisc*™ is also available on CD-ROM.

Science Abstracts is a valuable source of information that combines the disciplines of chemistry and engineering with those of physics, electronics, and computer science. Subject terms used in *Science Abstracts* are familiar and easy to use. *Science Abstracts* is also available online as INSPEC.

Science Citation Index (SCI) is an invaluable tool for researchers. It compiles the references cited in a subject area and lists the documents in which each citation occurred. *SCI* allows a researcher to trace the influence of an article and identify related work that would otherwise be impossible to discover. *SCI* is also available online in CD-ROM.

Applied Science and Technology Index provides broad coverage of English-language periodicals in science and engineering and is an excellent source for recent technical developments. It is available online and on CD-ROM.

ABI-Inform is an electronic database covering over 800 business and management periodicals including professional publications, academic journals, and trade and news magazines. *ABI-Inform* contains valuable information on companies and new product developments and is available online and on CD-ROM.

Electronic Databases

Electronic databases may be classified according to those accessible *online* and those distributed on *CD-ROM*.

Online (Remote Access). Advancements in computing and telecommunications have supported the remarkable proliferation of electronic information systems since the mid-1960s. Online searching is useful for the following reasons:

Complex concepts can be combined for a single set of results.

Results can be narrowed, reformatted, and downloaded to print or to store in a PC.

Materials not available locally can be accessed.

Recent and retrospective records can be accessed in one search.

It is important to remember, however, that online searching is not always the best approach to finding information. Material covered before the files appeared in machine readable format are not retrievable online. The search strategy must be planned in advance and based on knowledge of the online system and database structure being used. Search costs include charges for telecommunications, database access, and printed or downloaded records. These costs quickly add up.

Electronic databases allow for retrieval of many kinds of information including the following: chemical reaction files, physical property data, patent records, molecular formulas, full-text articles, buyer's guides, registry numbers, and chemical names. Structure searching with the user drawing or typing an alphanumeric representation of the chemical structure is nomenclature-independent and provides precise identification by using structural concepts.

Full-text databases offer the ultimate in comprehensive coverage. Files such as the American Chemical Society's *ACS Primary Journals*, the *Kirk-Othmer Encyclopedia of Chemical Technology*, and Mead Data Central's LEXPAT (patents) and NEXIS (news) allow the researcher to retrieve specific numerical data, techniques, or process information, and information from specified sources without having to select precise indexing terms. However, this tolerance of imprecision has its cost: increased access charges and online time, full screens that are difficult to read, and retrieval of irrelevant records. Full-text files often illustrate that less is still more.

CD-ROM Databases. Electronic databases are becoming increasingly available on CD-ROM for hands-on access through the use of a personal computer. Some database producers offer specialized subsets of their master files such an Engineering Information's *EiChemDisc*™ and the *McGraw-Hill Concise Encyclopedia of Science and Technology*. Files such as the *Kirk-Othmer Encyclopedia of Chemical Technology* provide a good starting place when you begin a new project or must deal with unfamiliar technologies in the laboratory.

The lack of standardized search software and most systems' inability to support many simultaneous searches are the main shortcomings of CD-ROM technology.

Patents

Patents are a source of valuable information ranging across new processes, products, and old applications for undeveloped discoveries that might have potential today. Patents may reveal what competitors are working on and contain significant information that appears nowhere else. In Europe and Japan, patent applications called "quick issues" are published, although U. S. patents are issued only after they have been granted. Both quick-issue and granted patents are worthwhile sources of information.

Patents are not easy to search. Organizational names and technical nomenclature vary. Also, in an effort to be precise, the authors of patents tend to use complex, convoluted terminology to describe even the most simple device or process. For these reasons, a good way to keep abreast of patent information is by scanning technical news magazines, technical journals, and review articles. Patents can also be searched online in *Chemical Abstracts* and *Derwent-World Patents Index*.

Copies of U. S. patents are available from Patent Depository Libraries across the U. S. and from the U. S. Patent Office in Arlington, Virginia. The addresses of major patent offices throughout the world and copy prices are listed in the Introduction to *Chemical Abstracts'* semi-annual volumes. Patent delivery services such as Chemical Abstracts, Inc. and IFI Plenum Data Company are faster but more expensive.

Reference Books

Among basic sources of important information for chemical engineers are reference books such as handbooks, directories, and encyclopedias. Typically, data are compiled from a variety of sources and provide a useful starting point for a search although they might not be definitive or error free. Because new editions of handbooks do not carry over all previous data, it is important to retain past editions and refer to them.

Below is a list of representative reference books that are especially useful for the chemical engineer:

Beilstein's Handbuch der Organischer Chemie (Handbook of Organic Chemistry)

The Chemical Formulary

CRC Handbook of Chemistry and Physics

Dictionary of Organic Compounds

Gmelin Handbuch der Anorganischer Chemie (Handbook of Inorganic Chemistry)

Kirk-Othmer Encyclopedia of Chemical Technology

Lange's Handbook of Chemistry

McGraw-Hill Encyclopedia of Science and Technology

Merck Index

Perry's Chemical Engineer's Handbook

Standards and Specifications

Standards are set by national and international organizations, professional and technical associations, corporations and government agencies. The National Institute of Science and Technology (NIST—formerly the National Bureau of Standards) collects standards used in the U. S., including codes of practice, rules for nomenclature and terminology, and guidelines for processes and procedures.

Many standards established by the American Society for Testing and Materials (ASTM) contain information relevant to chemical researchers. American National Standards Institute (ANSI) standards might also prove useful. Standards and specifications for most U. S. manufacturers are available from Information Handling Service on microfilm, CD-ROM, or online. Standards are also indexed in *Engineering Index* in print and in CD-ROM; standards are indexed online by *NIST Online*.

Manufacturing and supplier information is available in indexes such as the *Thomas Register* and subject-specific directories such as the following:

OPE Chemical Buyers Directory

Chem Sources—USA

Chem Sources—International

Gardner's Chemical Synonyms and Trade Names

Property Data

Property (chemical, physical, materials) data are scattered throughout the literature, and can be inconsistent and often unevaluated. Data from the National Standards Reference Data System (NSRDS) are an important exception. NSRDS data are critically evaluated before appearing in the *Journal of Physical and Chemical Reference Data*, co-published by the ACS and AIP for NIST, as well as in various technical journals.

Most collections of property data appear in handbooks, journals, bulletins, or abstracts and indexes that are compiled from other sources. Property data in trade journals are rarely indexed despite being extremely applicable to processing and handling operations. For this reason, it is particularly important to maintain a personal index of this information, including the sources for future reference.

Property sources, directories, and handbooks include the following:

> *CODATA Directory of Data Sources for Science and Technology*
>
> *Thermophysical Properties Research Literature Retrieval Guide 1900–1980*
>
> *Landolt-Bernstein Numerical Data and Functional Relationships in Science and Technology*
>
> *Corrosion Data Survey*
>
> *Thermophysical Properties of Matter*

Increasingly, property data are becoming accessible online. In addition, systems are being developed to interact with data input by the searcher. For instance, physical data for organic compounds are available in Beilstein online. In 1991, STN International began to offer the Materials Property Data Network consolidating a number of materials properties databases. Other properties databases include the Merck Index Online, Chapman and Hall (Heilbron) and the Corrosion Database. DIPPR is the computer-based system being developed by the Design Institute for Physical Property Data.

Conclusion to Part I

The information challenge in chemical engineering is not a lack of material but the vast quantity of potentially useful data and the degree to which it exists in widely scattered sources.[3] To establish a critical thinking frame of reference, each engineer must create a personal information frame of reference by becoming familiar with the available information resources. To maintain a critical thinking frame of reference, each chemical engineer must systematically preserve the results of all searches and remain open to new information retrieval technologies and techniques.

Part II: Broadening the Literature Search

Introduction to Part II: The Importance of Intellectual Breadth

The Office of Naval Research once conducted a study identifying critical factors related to technical creativity and effective performance. It found that successful researchers employed the following strategies: they routinely collected background information and ideas from existing literature, associates, and experts; they obtained information from sources not in their disciplines; they reviewed the literature for applicable past work; and they became the best-informed people in the work group.[4]

Researchers use current technical information to remain innovative and productive. They apply their awareness of recent findings to the development of new products and the improvement of old ones without duplicating work or infringing on the proprietary rights of others. Equally important, they use information to update their skills and to keep abreast of advances in technology. Part of the critical thinking model described in Chapter 1, the capacity of intellectual breadth is served if you keep current with research that is related to your field. This process is complementary to conducting research within your field, the subject of Chapter 3.

Current Awareness

Rapid advancements in chemical engineering and related fields are generating exponential growth in the technical literature, increased specialization, and a greater diversity in technical journals. Keeping up-to-date is further complicated by heightened interdisciplinarity as once-independent disciplines combine to generate new technologies. While this proliferation of subject specialities has opened publishing opportunities, it has also raised questions about overall quality of the journal literature. At the same time, the sharp rise in journal prices limits the number of subscriptions either a library or an engineer can afford.

More than ever, then, it is important to develop a personal strategy for keeping up-to-date by focusing on the most relevant sources and balancing the need for different kinds of information.

R. E. Maizell proposes a hierarchical approach to keeping current in technological fields such as chemical engineering.[5] Figure 5.1 below captures the approach:

Figure 5.1 A Plan for Keeping Current in Technological Fields

Maizell suggests regularly scanning a broad range of news magazines, a moderate number of general research journals, and a few specialized journals. As a student of chemical engineering, here are examples in each category that you may find helpful:

News Magazines:

Chapter One: The AIChE Magazine for Students

Chemical and Engineering News

Chemical Engineering

Chemical Engineering Progress

Chemical Engineering Education

Chemical Week

Chemistry and Industry

General Research Journals:

Harvard Business Review

High Technology

Journal for the Society of Technical Communication

Nature

Science

Technology and Culture

Specialized Journals:

AIChE Journal

Chemical Processing

Combustion and Science Technology

Hydrocarbon Processing

Industrial and Engineering Chemistry

Journal of Catalysis

Journal of Chemical Engineering Data

Journal of Physical Chemistry

Alerting Services

Alerting services, including contents page services and Selective Dissemination of Information (SDI) services provide broad access to the current journal literature.

Contents page services reproduce the tables of contents from a set of relevant journals. Subscribers can scan these pages and order full copies of selected articles. An advantage of this approach is that contents can be scanned quickly to see what work is being reported and by whom. Many corporate libraries produce contents page announcements bulletins of locally held journals. Copies of articles are easy to order and are in English. Commercial services include *Current Contents* published by the institute for Scientific Information (ISI) in broad areas of interest. Although the system for ordering copies of articles is easy to use, some journals may not be copyable and others are not in English. ISI's *Custom Contents* gives the subscriber more control by allowing for the personal selection of journals to be covered.

SDI services, offered by libraries, information brokers, and data base vendors, match the subscriber's interest profile to relevant databases and extract the most recent records for review. SDI has the advantage of providing abstracts and more focused results. Copies of the full article may also be requested.

Obtaining Assistance

The wide variety of information products, expansion of access systems, and development of new services make it more challenging than ever to find the right information in a timely and economical manner. Fortunately, there are places where the researcher can find help.

Personally visiting a library is always rewarding, particularly to consult with librarians or information specialists. These experts often provide answers more rapidly and can help the requester focus the search for more effective results. Corporate libraries have the advantage of being planned to meet the specific research needs of the corporation. Academic, research, or large public libraries often provide assistance to researchers who lack access to technical libraries where they work.

Information brokers are individuals or businesses that provide a wide variety of individualized services for their clients. For the researcher who lacks other access to technical information, the benefits of using a broker outweigh the costs. Brokers perform a variety of services including searches, document delivery, and current awareness.

Computer networks and gateways are not only improving communication among colleagues, but they are making it possible to deliver directly to a researcher's personal computer options such as downloading information into a word processing file, transferring structures drawn in a graphics program, and adding scanned spectra. It is also possible to store online searches and run them after hours at night rates.

Another option is to request reprints from authors, although this process takes time and response is sometimes uncertain. The reason for the request should be stated in a typed letter and accompanied by a self-addressed stamped envelope. It may be possible to establish a more collegial relationship by enclosing a copy of your own work.

Document Delivery

Copies of articles identified by a literature search can be obtained in a variety of ways. If the item is copyable it can be located and copied in a library. Some libraries will provide photocopy services free or for a fee; information brokers charge for document delivery services. Database producers such as *Chemical Abstracts* and *Engineering Index* offer document delivery to complement their search services.

Translations

Although there are foreign technical journals that are published in English, it might be necessary to obtain a translation of a desired document. Before contacting a translation service, however, there are some practical alternatives to pursue:

- Find an abstract in English to decide whether the needed information is present or whether the document merits continued interest.

- Seek books or review articles that describe the work.

- Ask the librarian if there is anyone in the university who might provide a translation.

Commercial translation services might be provided by free-lancers or by translation firms. Rough drafts cost less than high-quality versions, and translations from the Japanese are especially expensive.

Conclusion to Part II: The Interdisciplinary Chemical Engineer

In his book on finding chemical information, Maizell states the following:

> "Effective use of information helps avoid duplicating previously reported work. This achieves savings in time and funds and avoids infringing on the proprietary rights of others. In addition, even if there is no directly related previous work, [those] who make effective use of information can plan and act on a solid foundation of background data. Further, as a source of ideas or for idea development, chemical information sources and tools are invaluable fountains of inspiration and serendipity . . . [Those] who [know] how to use chemical information quickly and efficiently, and who [have] the required energy, imagination, and zeal, will usually have a clear advantage over the person who either lacks these skills and qualities or is too lazy to use them."[6]

For the student of chemical engineering, such advice on broadening information sources is especially important. In the future, Ralph Landau writes, there will be a need for an integrated chemical engineering education.[7] Thus, not only will research and scale-up become important, but so too will the ability "to lead a diverse team of specialists while retaining an overall economic marketing and technical perspective." To be able to address such market demands, you must be able to think critically. To perform such thinking will require the ability to keep current within and beyond the discipline of chemical engineering.

The next four chapters of this book—dealing with topics in communications, ethics, laboratory architecture, and laboratory reporting—will help you broaden your perspective.

Notes and References

1. For another article that provides guidelines for conducting a literature search, see Arthur A. Anthony, "The Literature: Becoming Part of It and Using It," In *The ACS Style Guide: A Manual for Authors and Editors*, ed. Janet S. Doss, Washington, DC: American Chemical Society, 1986.

2. R. E. Maizell, *How to Find Chemical Information*, 2nd ed., New York: Wiley, 1987, pp 6–8.

3. Chemical engineering is not alone in proliferation of information. For a review of the kinds of information retrieval needed in physics, chemistry, biology, geosciences, astronomy, engineering, mathematics, and computer science, see Constance C. Gould and Karla Pearce, *Information Needs in the Sciences: An Assessment,* Mountain View, California: The Research Libraries Group, Inc., 1991.

4. American Institute for Research, "Evaluating the Performance of Research Personnel," in R. E. Maizell, "Information Gathering Patterns and Creativity," *American Documentation,* (Vol. XI, 1960), pp. 9–17.

5. R. E. Maizell, *How to Find Chemical Information*, 2nd ed., New York: Wiley, 1987.

6. Maizell, p. 1.

7. Ralph Landau, "The Chemical Engineer and the CPI: Reading the Future from the Past," *Chemical Engineering Progress*, September 1989, pp. 25–39.

Five Assignments for Sophomores and Juniors: Understanding the Literature Search

Assignment 1

Recall the research planning guide from R. E. Maizell discussed in Part II. Draw a figure of your own process of performing a literature search.

Assignment 2

Do you have experience with electronic searching? Pick one experiment from your laboratory and run an electronic search on that experiment.

Assignment 3

Read Diana Crane's *Invisible Colleges: Diffusion of Knowledge in Scientific Communities* (Chicago: University of Chicago Press, 1972) and prepare a ten-minute oral report to your class on how information is exchanged in a scientific community. Then, spend five minutes of discussion time with your instructor to explore how Crane's findings are similar to the way that information is exchanged among chemical engineers.

Assignment 4

Recall Figure 5.1 and make a list of news magazines, general research journals, and specialized journals. Broaden this list by showing it to three of your chemical engineering instructors and asking for their additions.

Assignment 5

Identify two foreign language journals in chemical engineering. How are English language translations of these journals usually obtained?

Four Assignments for Juniors and Seniors: Understanding the Literature Search

Assignment 1

Identify your reference librarians and review with them the kinds of experiments you will perform in your chemical engineering laboratory course. Ask for guidance as to the most appropriate information retrieval source.

Assignment 2

Interview your laboratory instructor about what literature he or she reviews on an on-going basis. What advice can your instructor give you about keeping current in specialized topics of chemical engineering?

Assignment 3

Form an alerting services group with the members of your class. Decide what kinds of research will be most helpful in the laboratory and what information sources will be reviewed. To aid you in this effort, review Chapter 3 on collaborative work.

Assignment 4

Locate an information broker in your geographic area and find out the kinds of information retrieval the broker performs for clients in chemical engineering. Make a five-minute report to your class on the way that the broker operates.

A Great Books List for Chemical Engineers

In a series of lectures published in 1936, Robert Maynard Hutchins, a young chancellor of the University of Chicago, scathingly criticized American higher education. Hutchins argued that when fashion ruled, education declined. His solution? A return to the great books of civilization.

No list has yet emerged of great books for chemical engineers. *Within* the discipline, one would cite Walker, Lewis, and McAdams' *Principles of Chemical Engineering* (1923), Webber's *Thermodynamics for Chemical Engineers* (1939), Houghen and Watson's *Kinetics and Catalysis* (1947), and Bird, Stewart, and Lightfoot's *Transport Phenomena*. For its sheer utility, we might also cite Shreve's *Chemical Process Industries*, a standard since its first appearance in 1945.

But what about *beyond* the discipline? Following the motto of the University of Chicago—"Let knowledge grow from more to more, and thus be human life enriched"—we offer the following twelve books.

William F. Furter, ed. *History of Chemical Engineering*. 1980. Essays that provide an international perspective on the history of chemical engineering.

David A. Hounshell and John Kenley Smith, Jr. *Science and Corporate Strategy: Du Pont R & D, 1902–1980*. 1988. A splendid history about the process of research and design in one of America's leading chemical corporations.

Evelyn Fox Keller. *Reflections of Gender and Science*. 1985. A mathematical biophysicist argues that if science is not gender free, it will not necessarily be objective.

Bruno Latour and Steve Woolgar. *Laboratory Life: The Construction of Scientific Facts*. 1979, 1986. An anthropologist visits the laboratory and investigates how scientific order is constructed from an apparently disorderly nature.

(more)

Aldo Leopold. *A Sand County Almanac*. 1949. Elegant writing by a conservationist who believes that land is not merely soil.

Primo Levi. *The Periodic Table*. Translated by Raymond Rosenthal. 1984. "Every element," writes the Italian chemist and Holocaust survivor, "says something to someone (something different to each) like the mountain valleys or beaches visited in youth." A stunning book that reminds us that we are indeed what we do.

Walter Ong. *Orality and Literacy: The Technologizing of the Word*. 1982. Writing is a technology that restructures the way we think, Ong proposes. A compelling book that makes us re-think what we were really doing in all of those English courses.

James Rachels. *The Elements of Moral Philosophy*. 1986. An excellent introduction to moral philosophy.

Lauren B. Resnick. *Education and Learning to Think*. 1987. A sixty-two page monograph that lucidly explains higher-order cognition.

Terry S. Reynolds. *75 Years of Progress—A History of the American Institute of Chemical Engineers, 1908–1983*. 1983. A great history of a great professional organization.

Alfred North Whitehead. *Science and the Modern World*. 1925. A work that connects science and religion, the humanities and mathematics. Whitehead makes us wonder whether specialization should necessarily breed pedantry.

E. Bright Wilson. *An Introduction to Scientific Research*. 1952. A classic text by a chemistry professor on the principles, techniques, and procedures of research.

6 Ethical Decision Making in Chemical Engineering

Eric Katz
Assistant Professor of Philosophy

"Whether true or false, my opinion is that in the world of knowledge the idea of good appears last of all, and is seen only with an effort."

Plato

Preview

In Chapter 6 you'll have a chance to think about

- a framework for critical thinking about ethics

- ethical considerations such as safety and risk

- the unique ethical demands of technology

- defining morality

- ethical systems for decision making

- guidelines for the development of your own ethical perspective

"It is a good beginning of your art to offer your first-fruits to the poor, who will both thank you and God for it, which I see by your silence you seem to consent to."

Isaac Walton, *The Compleat Angler*

Beginning to Think Ethically

Ethical decision making is a complex subject. In order to acquire a proper understanding of the role of ethics in chemical engineering, you must be willing and able to adopt different viewpoints and alternative perspectives. You must use your imagination to look at problems in new and creative ways.

So far in the book you have been introduced to a number of new perspectives for critical thought. As an instructor in philosophy, I will present in this chapter a number of strategies for posing and answering ethical questions in chemical engineering.

To begin, I would like you to think about ethics in a different way. The traditional method of discussing ethics is simple, straightforward, and probably boring. There are specific rules, moral laws, codes of behavior, whatever, that we are all supposed to follow. In general, we all know what these rules and codes are, and to varying degrees we follow them. "Tell the truth." "Keep your promises." "Do not steal or cheat, even when you know you can get away with it." "Don't accept bribes." "Treat people with courtesy and respect." And on and on.

A discussion of ethics on this level usually disintegrates into a sermon or a lecture about following the rules. This is boring, as I said, because we generally know what these rules are without being told—they are not controversial—and by and large we follow them. We do not follow them all the time, for we make mistakes and we give in to temptation. We choose, in some circumstances, to ignore the ethical rules.

But no one in an upper-level chemical engineering laboratory course is prepared systematically to ignore or to break these basic moral principles. To do that would be to declare yourself unfit for a civilized life among other human beings; only psychopaths are systematically immoral.

So let us think about ethics in different ways, ways that are directly relevant to your future lives as chemical engineers. The crucial question I would like you to keep in mind is this: "What are the special or unique connections between ethics and engineering?" What do engineers do that creates a special class of ethical considerations?

Now once we ask this question, the answer seems pretty obvious. Engineers create technologies, or they use technologies to create products, artifacts, and commodities for human society. Some of these technological products are complex, some are simple; some of the technologies themselves are complex, and some simple. But just about everything we use in the contemporary world is designed by an engineer, and produced by a technology that is also designed by an engineer. Engineering and technological activity affect all aspects of human society.

This pervasive influence on human life creates a set of profound ethical obligations. Because the consequences of engineering work are felt everywhere, engineers have the moral obligation to evaluate the results of their activities.

To pursue the relationship between ethics and engineering more closely, we must examine the connection on two levels. We will address ourselves to this question: Where does ethical evaluation occur in the process of engineering?

Level 1: Safety and Risk

The first and most basic level of ethical evaluation in engineering involves the concepts of safety and risk. Indeed, we might want to say that a concern for safety and the reduction of risk is the primary ethical obligation of the engineer. Engineers are morally required to design products that are as safe as possible, within the limitations imposed by the state of current knowledge, the availability and characteristics of materials, and the economics of costs and benefits. Engineers are in a special position because they have more knowledge than other people about the qualities and characteristics of the products they design. It is that special extra knowledge that creates the ethical responsibility; the engineer knows better than anyone else the risks involved in the product he designs.

But focusing on a concern for safety is only a platitude. It is not enough. We all know that engineers are supposed to be conscious of risks. After all, every professional code of every engineering society lists a "concern for public safety" as being a leading professional responsibility of engineers.[1] Concern for safety begins to sound like one of those boring ethical rules I mentioned earlier: "Keep your promises, tell the truth, and design safe products." Is that all there is to ethics and engineering?

The problem with the ethical rule about safety—what makes it an interesting and important issue—is that engineers rarely work in isolation from non-engineering pressures. Most of the time, engineers work for profit-making corporations or for governments or universities that have goals which lie outside engineering safety. Engineers are routinely asked to balance—perhaps even compromise—their engineering standards with other criteria, such as cost, time-pressure, international or industrial competition, or political expediency.

Perhaps the most notorious recent example of this kind of conflict is the case of Roger Boisjoly, the engineer at Morton Thiokol in charge of the O-rings for the space shuttle booster rockets. Boisjoly recommended that the Challenger not be launched in cold temperatures because of the adverse effect of cold weather on the O-ring seals. But there was tremendous pressure from the corporate managers at Morton Thiokol and the government scientists at NASA to launch the Challenger as scheduled. Under these kinds of external pressures—that is, external to engineering practice—what power does a lone engineer have? The concern for safety will be crushed under the weight of other goals. The most chilling event in the entire Challenger episode was the statement made to Boisjoly's superior by the corporate managers: it is time, they said, to take off your engineering hat and put on your management hat. That statement, in a nutshell, summarizes the ethical dilemma faced by the engineer who is concerned with safety.[2]

Level 2: Ethical Characteristics of Technology

Once we see that engineers do not work in isolation from outside forces, we enter a second and deeper level in the ethical evaluation of engineering activity. And it is here that I ask you to stretch your imaginations to view engineering and technology from a wholly different perspective.

Rather than think about the ethical consequences of engineering or technological products, think about the ethical characteristics of the products themselves. I would like you to see that technology has an inherent ethical, social, and political dimension. Technological products are always the creation of broad social forces. Technological products are always designed to meet specific economic, social, or political goals. The simplest of these goals is the profit-motive of the independent business corporation. For instance, alar (daminozide) was produced since 1968 by Uniroyal Chemical Company to keep ripening apples on trees and to keep apples firm during storage. The scale-up agenda was economic. There can, of course, be more complicated goals, such as the production of penicillin to help the war effort in 1941. (For more about the history of penicillin production, see Chapter 2.) But whatever the goal, the overall point should be clear: technological products are created for social and political reasons. The political theorist of technology Langdon Winner summarizes this by saying that "artifacts have politics"—that is, technical products have, in themselves, political and ethical characteristics.[3]

I am asking you to grasp a rather subtle point. One way into this point is to consider the distinction between *making* and *using* a technological product. The traditional view is that making a product has no ethical dimension; it is only how the product is used that raises ethical issues. A knife can be used to slice bread or to slice someone's face; the user, not the maker, requires the ethical evaluation. But this traditional view ignores the fact that products are only made as a result of specific social forces, forces that carry with them ethical dimensions. There are always reasons for making a product, and these reasons can always be ethically evaluated.[4]

Winner's favorite example of the political and ethical dimensions of artifacts is one I experience everyday on my drive to work in Newark. I live in New York State, and I drive the New York State parkway system. The parkway system was designed by Robert Moses in the 1930s and 1940s. You may not realize this, but we call our highways (in the United States) "parkways" because Moses designed the New York State system as a series of roads to be driven to the parks. Now aside from their beauty as highways, the most striking feature of the New York State parkways is their underpasses. These underpasses are built quite low, with some clearances only eight and a half feet. Moses built them low for a specific set of reasons. He wanted to prevent trucks and buses from using the parkway system, and he particularly wanted to keep buses from driving to the state parks. Why was this? Because the buses would bring poor people from the city to his beautiful suburban parks; only middle-class people who could afford private cars were to be allowed easy access to the state park system.

Here then is a perfect example of the ethical dimension of a technological product. You would not normally think that highway underpasses would be ethically significant, but the ones in New York State were built at a certain height because the designer had a social and political agenda, a bias against the poor. In its creation, in its making, not merely in its use, the technological product is ethically significant.[5]

Let me extend the point and apply it to chemical engineering. To return to the example of alar, you may recall that in 1985 the U.S. Environmental Protection Agency (EPA) discovered that daminozide and a breakdown product (UDMH) might cause cancer. According to the EPA, UDMH could cause about 45 cancers per million people. From 1985 to 1987, Uniroyal's sales of alar dropped by about 75%. The alar controversy then became fueled by charges that the public and media overreacted—the alar "crisis" was treated as a leading public health threat—while the government underreacted—it did not ban alar immediately.[6] The alar case is striking in that it embodies a kind of shortsightedness typical of a technological innovation. A new technological entity—in this case, alar—is produced, but its original purpose, its consequences, and its impact—in short, its meaning for society is not fully considered. One result, then, is a failure to think through the ways that risk issues are communicated. It is almost as if, once the EPA study was completed, the work was finished. Because artifacts have politics, because technologies affect the structure of society, the work was just beginning.

What does all of this mean for engineers who wish to think about ethics, who wish to perform their jobs and pursue their profession with a high degree of ethical awareness? Engineers have to realize the corporate and social pressures that will, in part, determine the range of their activities.

Engineers have to realize that the design and creation of products and technologies do not occur in a social, economic, or political vacuum.

Let me summarize the discussion thus far. Engineering ethics is much more than following simple ethical rules. As is the case of the communications model provided in Chapter 9, engineering ethics requires an understanding of how engineering activity—the creation of technological products—is embedded in a social, economic, and political system. You have to look beyond the Scheibel Column and see the connection of technology to the outside world. You have to know more than whether or not a given design "works." You have to know what the technological product means for the rest of society. To be an ethical engineer is to be aware of the social and political consequences of technology.

Let me narrow the discussion now by turning to a definition of the concept behind ethical awareness: morality.

A Definition of Morality

In his short history of ethics, the philosopher Alasdair MacIntyre tells us:

> that in asking moral questions of a certain kind . . . we may discover that we cannot answer them until we have asked and answered certain philosophical questions.[7]

In practical terms this means that the moral rules that I discussed above—"Tell the truth," "Keep your promises," "Do not steal or cheat, even when you know you can get away with it," "Don't accept bribes," and "Treat people with courtesy and respect"—may prove worthless unless ethical issues are tied to a working knowledge of the history of ethical philosophy.

How does a chemical engineering student approach philosophical systems without becoming a philosophy major? To begin, you should develop what James Rachels terms a "minimum conception of morality."[8] Since there exists a history of rival philosophical systems of ethics, many of which are designed to contradict the others, you will become quickly confused unless you can develop a core definition of ethics.

Rachels' definition is especially helpful. He writes,

> Morality is, at the very least, the effort to guide one's conduct by reason—that is, to do what there are the best reasons for doing—while giving equal weight to the interests of each individual who will be affected by one's conduct.

Two points here are significant. The first is that conduct should be guided by reason, not by emotion. The second point is that the interests of others should be promoted. In other words, according to Rachels, to behave ethically you must become a "conscientious moral agent." He defines a conscientious moral agent as a person

> who is concerned impartially with the interests of everyone affected by what [you] do; who carefully sifts facts and examines their implications; who accepts principles of conduct after scrutinizing them to make sure they are sound; who is willing to "listen to reason" even when it means that [your] earlier convictions may have to be revised; and who, finally, is willing to act on the results of this deliberation.

Two Ethical Systems: Utilitarianism and Kantianism

With this definition of ethical behavior in mind, we may now turn to two philosophical traditions that will confront you as you form your own definitions of ethical behavior.

Utilitarianism. Perhaps the ethical system you are most likely to face in America is that developed by David Hume, Jeremy Bentham, and John Stuart Mill. Utilitarianism is a kind of "bottom-line" ethical theory: it is concerned with results. Utilitarianism attempts to measure the consequences of action, calculating the good and bad results, and thereby arriving at a mathematical formulation of the proper ethical choice.

The core principle of utilitarianism is found in Mill's Greatest Happiness Principle:

> The ultimate end, with reference to and for the sake of which all other things are desirable (whether we are considering our own good or that of other people), is an existence exempt as far as possible from pain, and as rich as possible in enjoyments.[9]

This principle is often referred to in a short-hand version as "the greatest happiness for the greatest number."

For anyone who deals with technology in an organizational institution or with political decisions in a large social unit, this ethical system seems promising. It appears to answer several problems about the determination of ethical goals and actions.

First, the objective character of a calculation, measuring the good and evil in a proposed course of action, has a great appeal to engineers and political decision-makers. Utilitarianism claims that a simple algorithm can be used to decide ethical problems: which alternative action produces the greatest ratio of good over evil results? Second, this ethical theory explains the role of happiness and its connection to the existence of pleasure and pain. Utilitarianism, in its basic version, simply equates happiness with an increase in qualitative pleasures: thus the pursuit of pleasure and the avoidance of pain become the goals of ethical action. A third advantage of utilitarianism is that it seems to deal with responsibilities toward others: the calculation of pleasure and pain that is the basis of the ethical decision is not focused on the individual, but on the social unit as a whole. After all, utilitarianism is concerned with the greatest happiness for the greatest number.

Unfortunately, however, the utilitarian system has a fundamental weakness: utilitarianism emphasizes the overall consequences of an action for the majority instead of the rights of the individual. The happiness of one person, or the happiness of a minority of the population, may be sacrificed for the total happiness of the overall population. The good consequences of alar, for instance, may override the fact that 45 cancers may be caused per million people exposed to alar over their lifetime—thus the greatest overall happiness is achieved, but only by violating the rights, increasing the pain, of a minority of the population—those individuals who contract cancer due to the use of alar in our society. Because utilitarianism emphasizes total or overall good, it may set aside too easily the interests or rights of specific individuals.

Kantianism. If you are to extend your ability to think critically beyond utilitarianism's appeal to the good consequences for the majority, you may do so by examining the philosophy of Immanuel Kant.

In his *Grounding for the Metaphysics of Morals* (1785) Kant set forth his supreme moral principle, what he called the "categorical imperative":[10]

Act only according to that maxim by which you can at the same time will that it should become a universal law.

What this means is that each person should make moral decisions based on the possibility that the decision would be a universal law, binding on all other people, without a logical contradiction. If a person considered murdering his chemical engineering lab instructor, he would first consider whether "murder" could become a universal law. But a law of universal murder would lead to the death of everyone, including would-be murderers—this is a logical contradiction, and thus murder is an immoral action. Kant thought that in practical terms the categorical imperative meant that all other people should be treated with respect, as an individual with worthy goals and aspirations, as an "end in itself." No moral person would ever treat another person merely as a means to his own satisfaction; no moral person would ever use another person as a mere object.[11]

There are two benefits to an ethical system informed by the Kantian categorical imperative. First, the individual gains a force not allowed by utilitarianism. In fact, the very power of ethical force rests not with a total system but with an individual. The notion of the greatest good is scrapped in favor of the idea of universal rules based on reason—a reason accessible to and derived from individual thought about moral consistency and respect. Kantian ethics stresses the individual's central role in the ethical decision-making process; because of this role, individual rights cannot be overridden for the good of the total social unit. Second, the use of the categorical imperative removes any hierarchies of ethical value. In a Kantian ethical system, all beings worthy of respect are equal; individuals must put themselves on a level with all other individuals. The central idea behind the universal binding law of the categorical imperative is that any valid moral principle applies to everyone: there are no exceptions to be made on the basis of power or status.

As this discussion reveals, you do not have to major in philosophy to understand philosophical questions. You do, however, have to be aware that (1) ethical thought demands rationality, and (2) ethical thought may be broadened by a study of ethical systems.

Guidelines for Developing Your Own Ethical Perspective

In this chapter I have asked you to think about ethics in a new way, one that will help you develop your beliefs before you will have to put them into action. In order to begin to broaden your ethical perspective, I suggest that you use the guidelines below:

Guideline 1. Ethical awareness must be articulated in a variety of circumstances if it is to be effective. You must continually examine your ethical ideals in practical case studies. For example, a survey on ethical choice in a 1991 issue of *Chemical Engineering Progress* asked the following: What would you do if you found that management knew that five stuffanol reactors at your plant were regularly operating at 180 degrees centigrade although the limit switches were supposed to be set at 175 degrees centigrade? Would you do nothing, tell the safety inspector, try to persuade your boss with technical data that the reactors were being neglected, go over your boss's head to senior management, or look for another job?[12]

Test cases such as this one become essential if you are to develop an ethical perspective. With your colleagues in your laboratory course, you should discuss how you would react in various situations and the reasons for your point of view.

Guideline 2. Ethical responsibility implies neither whistleblowing nor silence; it does imply the rational evaluation of alternative actions. This point is made by Caroline Whitbeck in her analysis of Roger Boisjoly's predicament.[13] As the Challenger explosion reveals, the issues involved in a case of technological impact are often far too complex for extreme or emotional reactions. Instead, rationality should win the day. A calm and measured discussion of the issues is the best method of resolving ethical problems. In the case of alar discussed above, it is clear that a better method of risk communication would have given the public more useful information and avoided the outcry that followed when alar was found in apple juice, the drink of choice by parents for little children.

Guideline 3. Technical concerns are tied to humanistic concerns. There is no point in trying to dissociate technology from the humans who made it. Technological artifacts, as I stated above, have politics; the social and political purposes of a technology should always be kept in mind. When we focus on the purposes of a new technology, it is easy to see that ethical problems may be avoided by simply defining technical specifications. For example, in the Challenger explosion discussed above, it is clear that low temperatures caused the O-rings to fail to seal. However, the contractor seems to have had no technical criteria for the lowest possible temperature acceptable for launch conditions. Thus, the ethical considerations were tied to technical considerations. In failing to see this relationship, the contractor became responsible for what Boisjoly predicted would be "a catastrophe of the highest order—loss of human life."

Guideline 4. To remain ethically alert, you must consider the power of bureaucratic organizations. It is difficult to maintain individual thinking and action within large scale institutions such as corporations, the civil service, or universities. Over time and circumstance, individuals tire. When they do, their decisions become routine. In the survey discussed above in Guideline 1, the individual who faced the ethical choice also faced a supervisor who had been in the plant for thirty-five years, was about to retire, and had never been known to initiate major changes. Would change be likely in that situation? To be ethically alert, then, you must ask yourself these questions: How can I remain awake? How can I avoid the temptation of routine? How can I maintain my own personal standards, my own ethical perspective?

Guideline 5. The ability to communicate effectively is central to the maintenance of an ethical position. As will be discussed in Chapter 9, without effective writing and speaking abilities, the engineer can never function fully as a professional. With these abilities, an engineer is able to control—not be controlled by—the working environment.

Notes and References

1. The Code for the National Society of Professional Engineers (NSPE) lists the following as its first "fundamental canon": "Hold paramount the safety, health and welfare of the public in the performance of their professional duties." Similarly, the Accreditation Board for Engineering and Technology (ABET) lists as its first principle, "Engineers shall hold paramount the safety, health, and welfare of the public in the performance of their professional duties." For more on engineering codes, their use, and limitations in the determination of ethics, see the following: Stephen H. Unger, *Controlling Technology: Ethics and the Responsible Engineer*, New York: Holt, Rinehart, and Winston, 1982, pp. 32–55; Mike W. Martin and Roland Schinzinger, *Ethics in Engineering*, 2nd ed., New York: McGraw-Hill, 1989, pp. 86–92. Both of these texts contain several engineering codes.

2. Boisjoly reports this discussion in a 1987 address he gave to engineering students at the Massachusetts Institute of Technology. The text of the speech can be found in Caroline Whitbeck. "The Challenger Disaster," *Books and Religion,* 15 (1987), pp. 3–4, 12, 28. The text is also reprinted in Deborah G. Johnson, ed., *Ethical Issues in Engineering,* Englewood Cliffs: Prentice Hall, 1991, pp. 6–14. For more discussion see Norbert Elliot, Eric Katz, and Robert Lynch, "The Challenger

Tragedy: A Case Study in Organizational Communication and Professional Ethics," *Business and Professional Ethics Journal,* 12 (1993). Forthcoming.

3. Langdon Winner, *The Whale and the Reactor: A Search For Limits in an Age of High Technology,* Chicago: University of Chicago, 1986, pp. 3–39. For an interesting treatment of the chemical engineering curriculum along the lines of Winner's hypothesis, see M. V. Sussman, "Engineering Schools Train Social Revolutionaries! Isn't it Time Our Students Were Told," *Chemical Engineering Education,* (Spring 1987), pp. 78–80.

4. Winner uses the knife example to illustrate the distinction between making and using and the traditional view of technology discussed above.

5. Winner, pp. 22–24.

6. A. M. Thayer, "Alar Controversy Mirrors Differences in Risk Perceptions," *Chemical and Engineering News,* (August 28, 1989), pp. 7–14.

7. Alasdair MacIntyre, *A Short History of Ethics,* New York: Macmillan, 1966, p. 5.

8. James Rachels, *The Elements of Moral Philosophy,* Philadelphia: Temple University Press, 1986. The definitions of "morality" and "conscientious moral agent" can be found on p. 11.

9. John Stuart Mill, *Utilitarianism,* Indianapolis: Hackett, 1979, p. 11. Originally published in London in 1863. Mill's work was an elaboration of the thought of Jeremy Bentham, *An Introduction to the Principles of Morals and Legislation,* 1789.

10. Immanuel Kant, *Grounding for the Metaphysics of Morals,* Indianapolis: Hackett, 1981, p. 30.

11. Kant, p. 36.

12. Cynthia F. Mascone, Alex G. Santaquilani, and Charles Butcher, "Engineering Ethics: What are the Right Choices?" *Chemical Engineering Progress,* (April 1991), pp. 61–64. For responses of over 970 readers, see Cynthia F. Mascone, Alex G. Santaqualiani, and Charles Butcher, "Engineering Ethics: How ChE's Respond," *Chemical Engineering Progress,* (October, 1991), pp. 73–82.

13. Whitbeck, Caroline, "The Engineer's Responsibility for Safety: Integrating Ethics Teaching Into Courses in Engineering Design." Paper prepared for the National Science Foundation. #83–10751. For more on "whistleblowing" see Richard T. DeGeorge, "Ethical Responsibilities of Engineers in Large Organizations: The Pinto Case," *Business and Professional Ethics Journal,* (Fall 1981), pp. 1–14. Both Unger and Martin and Schinzinger also discuss whistleblowing; Unger, pp. 8–29 and 121–140; Martin and Schinzinger, pp. 203–229.

Assignment for Sophomores and Juniors: Developing Your Own Ethical Perspective

Throughout this book you have been given various systems for asking subtle questions about your field. With the discussion of the two philosophical systems described above in mind, what kinds of questions might you ask to strengthen your awareness of your own ethical perspective?

In *Science and the Structure of Ethics* Abraham Edel provides a framework for the analysis of ethical perspectives.[1] Here we adapt Edel's framework so that you may develop a set of questions which will help you define, articulate, and refine your own ethical perspective.

In order to help you think about your own ethical perspectives—both developed and emerging—write a paper which addresses each of the questions below:

1. What is your personal view of your field of study? What is Chemical Engineering all about? You might consider the ways that your field seems to function and advance. Answering this question will help you set the context for your ideas.

2. How does the individual function and act in your field? You might consider here the situations that are often encountered by chemical engineers. What professional and personal problems arise in the daily activities and careers of chemical engineers? How do individuals cope with these unavoidable problems?

3. How do individuals work in your field at specific institutional sites? You might consider and compare how people in your field work within governmental or private agencies as opposed to private business corporations. What kind of problems do they face within organizations? Do these problems have an ethical component?

4. What are the dominant dangers in your field? That is, what areas have proven most knotty for chemical engineers in the past? Are these merely technical engineering problems, or do they concern social and ethical issues?

5. How are these dominant dangers solved?

6. At present, which ethical systems seem most promising to you in helping you anticipate and solve problems in your field?

(more)

121

Assignment for Sophomores and Juniors: Developing Your Own Ethical Perspective

In answering these six questions, do not be afraid to admit that you have not considered—or do not know about—certain issues. After all, this exercise is designed to help you develop and refine your ethical awareness.

1. Abraham Edel. *Science and the Structure of Ethics*. Chicago: University of Chicago Press, 1961.

Assignment for Juniors and Seniors: Laboratory Ethics and Outlying Data Points

Perhaps you have begun to consider, and perhaps develop, a personal system of ethics and morality in engineering and technology which you will use during your professional career. As part of that development, the following exercise will focus on some issues which you will face in the chemical engineering laboratory. Your response to these will assist you in your ethical development.

The primary ethical issue facing you in the laboratory is the handling of data. This occurs on two levels: a) data collection and recording, and b) data analysis and comparison to a theoretical model or literature empiricism. Consider the first level below.

Because the lab notebook becomes a de facto legal document, whatever is written in it must be a truthful and objective description of the data. In addition, your ideas and observations may be entered. The best current example of the importance of the lab notebook is the "Baltimore Case".[1] In that case, it was claimed that entering incorrect or illegitimate data is not a significant act because the lab notebook is irrelevant to the established truth of a scientific hypothesis. But this action is unbridled intellectual arrogance and totally unethical. The behavior has endangered at least one career (a whistleblower), tarnished several others, and dragged the name of a prestigious institution through the muddy press.

Your analysis of your data, and its comparison to a theoretical relationship (or literature correlation), will challenge you ethically because you are functioning in a research environment. As described elsewhere in this book, the critical thinking environment of the laboratory requires you to integrate all your engineering and science knowledge, and to recognize the limits of your data.

Since you often do not have the time to collect a statistically significant amount of data, you must estimate the uncertainty limits of your data and calculated results. (See Chapter 4.) If your particular theoretical model (or the literature correlation) predicts a behavior which is not directly apparent in your data, what is the ethical thing to do? Consider the following example:

(more)

Assignment for Juniors and Seniors: Laboratory Ethics and Outlying Data Points

You have performed a kinetics experiment, obtaining concentration versus time data in a batch reactor. You wish to obtain the reaction order. First you write

$$\text{Rate} = -dC/dt = k\,C^n$$

where C = concentration, t = time, k = rate constant, and n = order. You plot C versus time, differentiate the curve to get rates, then plot $\ln(-dC/dt)$ versus $\ln C$ because

$$\ln(-dC/dt) = n\,\ln C + \ln k.$$

If the data in this plot fall in a straight line, the slope will be equal to n. You also plot appropriate uncertainty bars around each point. (Secretly, you are hoping for n = 1 so that you can claim fairly uncomplicated unimolecular chemistry.) You are disappointed because, not only is the slope not 1, but the line is curved. What should you do? Discuss the ethics of each option below.

1. You ignore the curvature and remove the uncertainty bars. You draw the best fit line of slope = 1, then claim that the equipment was not working well.

2. You go back to your data, and you "cook" them. The result is a much better behaving curve (i.e., nearly straight, slope nearly 1).

3. You go back to your data, examining it carefully for any possible systematic errors in the experimental procedure and/or your work-up.

4. Finding no errors, you have no choice but to accept the validity of the data. You re-examine your theoretical model to see if it is appropriate for describing the experiment which you performed.

5. Satisfied with your data and your theory, you plot the predicted curve and realize that it falls within the uncertainty bars of all but one data point. You reject that outlying point, and present your results.

6. If your chosen action is not listed here, what would you do? What are the ethical and moral implications of your decision? What are the reasons for your decision?

(more)

124

Assignment for Juniors and Seniors:
Laboratory Ethics and Outlying Data Points

You will be faced with this kind of issue many times in the laboratory course and in your professional career. How you respond to the above exercise may help you to stop and think about real-life ethical decisions in your profession. [2]

1. For a comprehensive treatment of the Baltimore Case, see Philip Weiss. "Conduct Unbecoming." *The New York Time Magazine.* October 29, 1989. pp. 40–41, 68–71, 95. For a review in *Chemistry and Engineering News*, see May 21, 1990, pp. 5–7.

2. For an interesting article on a chemical engineering course in ethics, see James C. Watters and Dominic A. Zoller. "Developing a Course in Chemical Engineering Ethics: One Classes' Experiences." *Chemical Engineering Education.* Spring 1991. pp. 68–73.

The Right Choice: A Survey on Engineering Ethics

In April of 1991, *Chemical Engineering Progress* conducted a survey unprecedented in its originality and scope. The editors constructed a survey around two case studies and tracked the combined responses of 1,300 British readers (who wrote to *The Chemical Engineer [TCE]*) and American readers (who wrote to *Chemical Engineering Progress [CEP]*). Because this survey is the first of its kind, it deserves our attention.

Scenario # 1: Margaret's Choice

Margaret was the chief process engineer at a 10-million-lb/yr facility that produces pigments and dyes. Recently, she had been given responsibility for the plant's environmental compliance. As she shouldered her duties, she began to suspect that the plant might be responsible for groundwater contamination. She can

A. Do nothing about the potential contamination.

B. Anonymously tip off the regulatory authorities.

C. Recommend to her boss that the plant conduct an internal review to determine if there is contamination and, if so, its extent.

D. Request a meeting with representatives of the corporate legal and environmental affairs departments, as well as the plant manager and the corporate director of engineering.

E. Look for another job while ignoring the potential problem.

F. Do something else.

Analysis of Scenario # 1

Only a few readers would take no action or immediately call the regulatory authorities. An overwhelming majority of the respondents (73%) said that they would ask the plant manager to initiate an internal investigation (option C).

The next most popular choice was option D. Fifteen percent of the respondents said they would request a meeting with the corporate legal and environmental affairs department, as well as the plant manager and the corporate director of engineering.

(more)

About 10% of the respondents wrote that they would be prepared to use a multipronged attack and that their choice represented only the first step that they would take.

Would the respondents in reality follow the choice they selected? Fifteen percent said that they would not. "I hope I would in reality find the courage to do the right thing," one anonymous respondent wrote. "In the past I have knowingly taken great personal flack for doing what I thought was right even though I knew the situation was unlikely to improve. However, now I have a family who would suffer along with me."

Is the right thing the most effective? Twenty-one percent of American readers (*CEP*) and 30% of British readers (*TCE*) responded that what was right was not necessarily most effective.

Scenario # 2: Tom's Runaway Reaction

Tom recently accepted a promotion in a plant, but he noticed that there was a potential for runaway reactions in his area. The limit switches on the reactors were unreliable, and Tom also suspected that operators were lifting the pens when temperature peaks would approach so that production would not be inhibited. Tom proposed better instrumentation but was turned down by the plant manager and the business manager. Eventually, an accident did happen, although no one was injured. The operator on watch was likely to get fired or demoted by the plant manager, but Tom felt this action was unfair and ineffective. Tom can

A. Do nothing.

B. Tell the safety inspector the true story.

C. Try to persuade Dick (the plant manager) with more technical data.

D. Go over Dick's head and talk to Henry (the vice president).

E. Look for another job.

F. Do something else.

(more)

Analysis of Scenario # 2

Only 3% of the respondents said that they would do nothing (option A) or look for another job (option E). Only 10% of the respondents said that they would tell the safety inspector (option B), although 20% of the respondents said they felt this was the right thing to do.

Forty-three percent of the respondents said they would try to persuade Dick with more technical data (option C). Option D— go over Dick's head—was far less popular. Only 23% of the respondents selected option D.

Is the right thing the most effective? Thirty percent of *CEP* respondents and 41% of *TCE* readers said that the right thing and the most effective thing are not the same.

And So?

We have only provided a thumbnail sketch of this extensive survey, and we refer readers to the April and October issues of *CEP* for more detail.

Two inferences seem clear. First, the overwhelming solution for both Margaret's environmental problem and Tom's safety dilemma, readers said, was to gather more data. This choice of action, as the editor's recognize, is "consistent with the kind of training that engineers get." Engineers are trained to make technical conclusions based on data and information, not on hunches. This training appears to carry over into the realm of ethical decision making. Second, the decision to do what is ethical is not necessarily congruent with what is most effective. "Might we," the editors ask, "in our normal way of doing business, be going about doing things the wrong way? Why aren't the most effective actions the most ethically correct?"

Indeed. We postulate the following. While engineers are correctly trained to gather information before making any type of decision, they are incorrectly trained in strategies by which ethical action and effective management may be made one and the same.

True, this alignment is a large order, but so, in the beginning, was the job of defining the profession itself. If any profession is likely to discover methods of uniting ethics and economy, that profession is chemical engineering.

7 Planning the Laboratory Environment: An Architectural View

Chris Cowansage, AIA
Director of Laboratory Planning
Discovery Center
CRSS Architects

"The New Yorker" March 9, 1992.

"A picture or poem is often little more than a feeble utterance of man's admiration out of himself; but architecture approaches more to a creation of his own, born of his necessities, and expressive of his nature."

John Ruskin, *The Stones of Venice* (1853)

Preview

In Chapter 7, you'll have a chance to think about

- how a community of laboratory users can make a meaningful contribution to the planning and design of a new laboratory facility

- how you would alter your present laboratory facility

"I shall commend any Angler that tries conclusions, and is industrious to improve the Art."

Isaac Walton, *The Compleat Angler*

Initiating the Laboratory Planning Process

After you graduate with your science or engineering degree, you may start your career working in a traditional laboratory setting which you might simply take for granted. At first, you might be relatively unaware of the profound impact that your research environment will have on your productivity, your health and security, and your morale and creative potential.

Over time, however, you will begin to discover the merits as well as the limitations of your facility. Through the constant use of the laboratory, often at odd hours or in the heat of a deadline, you will come to notice the significance of details in your physical environment. You will find yourself actually forming expert opinions on issues of laboratory planning.[1] Consequently, when you agree to participate in the design of a new laboratory, you will likely make a valuable contribution to the creation of a functional, safe, and aesthetic working environment.

Since a new laboratory, to be truly successful, must meet the particular design objectives established by those who will directly use the facility, it is important to convert all observations about your current laboratory into rational goals for the proposed laboratory. You can contribute to this effort in the following ways:

- Evaluate in an informal "lab notebook" the function, safety, and aesthetics of your current laboratory, including comments on layout of benches, instruments, equipment, and fume hoods, desk and office sizes and locations, as well as access to shared support, such as sinks, compressed air, etc.

- Summarize in a brief report the overall success of your current laboratory as a center for experimentation, procedural work, data analysis, benchside and lab-office meetings, informal discussion, telephoning, writing, and reading of reports and protocols.

- Isolate and openly discuss the key points which need to be addressed in the planning of the new laboratory once the design process gets underway.

A careful analysis of laboratory planning and design issues by the users of the laboratory can result in an environment precisely tailored to the demands of research, such as the one in Figure 7.1.

Analyzing the Laboratory

It is easy to forget, in the initial stages of planning, that a laboratory space is more than just a room. An initial review should reveal that a laboratory design will have to reconcile the key attributes of function, safety, and aesthetics. The result should be an interactive environment, driven by technology but serving human needs.

Function. The functioning of a laboratory depends, first of all, on the appropriateness of the interior layout. For example, casework (benches, shelving, and cabinets) must be selected in the appropriate types and lengths. Equipment must be designated as either bench or floor-mounted, depending on project needs. Fume hoods, likewise, are available in different types and sizes.

Figure 7.1 Computer Generated Overview of a Planned Laboratory (Courtesy CRSS)

Large hoods in chemistry labs may be installed parallel to laboratory walls, as in Figure 7.2 below, with the center left open and reserved for laboratory benches or equipment racks and tables. Multiple smaller hoods may be selected, on the other hand, when the aim is to accommodate a variety of research demands within a single laboratory.

Flexibility of layout is also critical to the functioning of the laboratory. Movable racks or tables, for instance, provide flexibility and prevent the future occurrence of ad hoc arrangements, such as the balancing of new, heavy, or oversized instruments on outdated, narrow benches.

Flexibility should also be considered regarding servicing of mechanical systems, so as to ensure minimal disruption of laboratory work. The Salk Institute for Biological Studies in La Jolla, California, designed by Louis Kahn in 1959, is a prototype for this type of functionalism. As shown in Figure 7.3 below, it provides a very high, accessible ceiling-zone, called an interstitial space, that enables the checking, repairing, or changing of plumbing, electrical or mechanical services without significant inconvenience to the laboratory personnel below.[2]

Flexibility is also important in the allocation of space. Space needs may shift as a result of accommodating administrative changes. In the example shown in Figure 7.2, the door between the laboratories promotes some degree of interaction among the occupants. However, should project needs change, the "double" laboratory can become a larger "single" laboratory through removal of the middle partition to provide a more continuous, interactive workspace.

Administrative flexibility is enhanced too when offices are located outside of the laboratories along a personnel corridor, allowing them to be assigned independently of the laboratories. An

Corridor

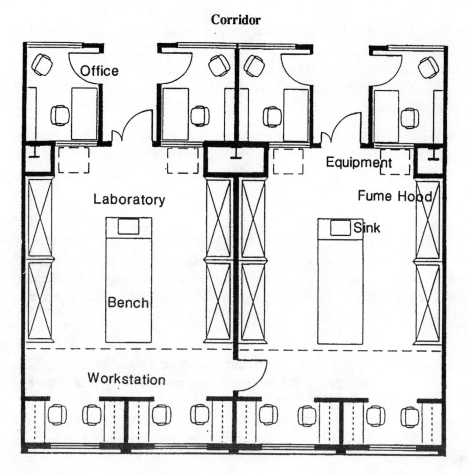

Figure 7.2 Laboratory Plan Showing both In-Lab Work Areas and Out-of-Lab Offices (Courtesy CRSS)

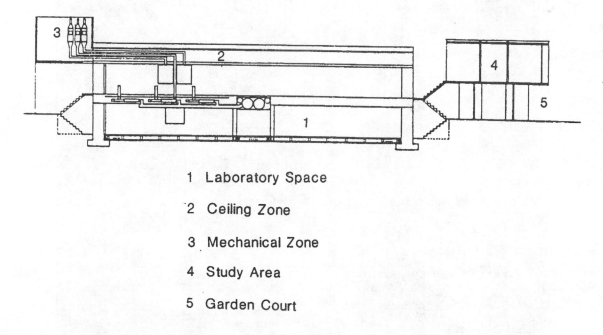

1 Laboratory Space

2 Ceiling Zone

3 Mechanical Zone

4 Study Area

5 Garden Court

Figure 7.3 Building Section Through Salk Institute (Diagram SRI.39 from *L.I. Kahn, Complete Works 1935–1974*. Heinz Ronner and Sharad Jhaven. Birkhauser: Boston, 1987)

early example of designed flexibility in laboratory architecture occurs with the two-person labs designed by Serge Chermayeff[3] in the 1930s. Chermayeff used a repetitive modular organization for all laboratory spaces, each with its own structural, electrical and plumbing functions, linked by a common circulation spine. This rational planning provides the infrastructure for a variety of laboratory configurations.

Today's laboratory buildings often require a mechanical infrastructure that can allow for the rearrangement of casework, fume hoods, equipment, and instrument space to satisfy changing project needs. Figure 7.4 below shows the mechanical infrastructure required to support a modern laboratory.

Safety. The safety of the laboratory is based in part on strict adherence to pertinent building and life safety codes. It is also a response to the particular hazards of the laboratory in question. Some of the critical responsibilities of the architect include providing adequate circulation and exits from a lab, separating hazardous procedures from nonhazardous ones, and providing safety equipment such as eyewash, showers, and extinguishers in appropriate locations.

In the laboratory design in Figure 7.2, there is always a second route out of the lab from any location, either into a corridor or into another lab; thus, if an explosion occurs in a fume hood, no one will be trapped in a dead-end aisle. No passive activities such as reading or writing at a desk occur near a hood, which can be a source of fumes or fire. Finally, there is an eyewash station at the sink and a safety shower near the laboratory exit, both in easy reach in case of emergency.

Figure 7.4 View of Building Systems: Utility Corridor (Courtesy CRSS)

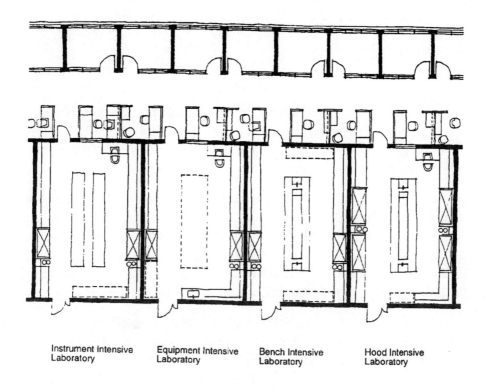

Figure 7.5 Plan of a Partial Laboratory Wing (Courtesy CRSS)

Aesthetics. The effectiveness of a laboratory can be measured by the degree of attention to convenience, comfort, and theories of aesthetics. While it goes without saying that to be conducive to productivity, a laboratory should provide adequate area per occupant of lab space, bench space, equipment, and desk space, these components should also be attractive. Inappropriate materials and colors selected for a laboratory can make it appear crowded and over-active once it is stocked with instruments and glassware. Choices of an aesthetic nature must be based on an analysis of potential optical and psychological effects. Harmonious materials and colors are important in a laboratory, which can appear very busy when fitted out.

The availability of direct, natural light and view for laboratory offices and indirect, natural light for laboratories is also generally desirable. An exception is an optical laboratory, wherein background light must generally be minimized. Offices outside of a laboratory, along a corridor at the building perimeter, benefit from both light and view. High windows or windows from perimeter offices can yield indirect light for the laboratory.

A sense of privacy is also a crucial feature for some workers and some types of work. At the same time, however, access to shared personnel support spaces—such as a library or reading room, a mail room, a small break area, a conference or training room, and a computer terminal room—can greatly enhance the quality of the lab building by addressing human needs outside of the laboratory.

Analysis of needs according to function, safety, and aesthetics results in a modern, well designed laboratory, as pictured in Figure 7.5.

The Laboratory Design Process

The laboratory design process ultimately requires a collective effort that supersedes individual contributions. Having applied your experience to the evaluation of many issues, including function, safety, and aesthetics, you and your laboratory colleagues must now engage in a joint design effort.

First, you must assemble a team which represents a multiplicity of perspectives on laboratory use. It must be able to consider the requirements of the lab from the points of view of, for example, project managers and administrators, senior laboratory investigators and laboratory technicians, as well as engineers and architects.

Second, your team must set in motion a comprehensive dialogue on the proposed laboratory, which elicits intellectual input from everyone who will ultimately use the facility. One way of inspiring a productive dialogue about how the lab environment can best serve the lab staff is by distributing a written probe to all, as illustrated in Figure 7.6. By studying the results, including the variations in responses, your team should be able to establish the fundamental goals of the design, respecting the project's mission, scope, and key planning issues.

Third, a critical step in achieving an integrated lab design involves the detailed visualization of verbal data about the current research environment. Expectations for the proposed laboratory, even those relating to casework details, piped services or electrical outlet locations, can be sketched in three dimensions, with views from above or within the lab, to illuminate the subtle forces that make the work environment satisfactory. The use of computer modeling for such visualization is an important means of clarifying and coordinating laboratory planning issues. When you and your colleagues actually see on paper what you have proposed, modifications might be needed.

What issues are key to accommodating the future needs of science in laboratory buildings?

	Yes	Perhaps	No
Laboratory Space			
More aisle space between casework	☐	☐	☐
Fixed in-place casework in the lab	☐	☐	☐
Deeper bench tops for instruments	☐	☐	☐
More racked equipment space in the lab	☐	☐	☐
More space for team work	☐	☐	☐
More lockable storage space in the lab	☐	☐	☐
Better access to cable trays/utilities	☐	☐	☐
Access to operable windows in the lab	☐	☐	☐
Office Space			
Larger faculty offices	☐	☐	☐
More office space for computers	☐	☐	☐
More student desk area in the lab	☐	☐	☐
More student desk area out of the lab	☐	☐	☐
Faculty office space nearer to laboratory	☐	☐	☐
More private office areas	☐	☐	☐
Meeting Space			
More team/work rooms	☐	☐	☐
More informal meeting areas for team/work	☐	☐	☐
More area in the office for meetings	☐	☐	☐
More quiet/private work area	☐	☐	☐
Computer equipped conference rooms	☐	☐	☐
Teleconferencing Capabilities	☐	☐	☐

Figure 7.6 Sample Probe for Laboratory Users (Courtesy CRSS)

Fourth, decision-making must be founded on objective conclusions supported by informed and reasoned judgments. Decisions arising from unstructured thinking, impulse, or blind adherence to precedent are irresponsible where large amounts of time and money are at stake. The design of an effective research facility occurs when you who will use the facility rely on hard evidence as the basis for resolving the key planning and design issues.

Fifth, your team ultimately needs to arrive at a plan which reconciles the broad underpinnings of the project with its narrow, day-to-day requirements. By harmonizing conceptual and pragmatic issues, the designer, with your team, imbues a built product with meaning. Several types of broad but very strategic issues typically arise early in the design process.

One of these issues relates to the long range research goals of the project. Will, for example, the building facilitate discovery or encourage innovation by fostering interdisciplinary interaction? Further, will the building be designed to accommodate changes in research focus and mission, or simply changes in technology, function, or staffing. Another long-range issue relates to business objectives. Will the building be designed to enhance productivity? Will the building contribute to recruitment and retention of valued staff? Finally, will efficient engineering systems and materials minimize the costs of operating the building? There are also important philosophical issues related to the design process. Will the building arise out of an original or ideological architectural statement? Or will the building defer to traditional values and historical prototypes that center around the existing spatial context? These broad considerations, and many others, come into play at every stage of the design process and have a profound impact on the ultimate outcome of the project.

Final Thoughts

The key issues of laboratory design must be thoughtfully resolved to realize a thriving laboratory building. The laboratory design illustrated in Figures 7.7 and 7.8 offer one solution. Many other creative options are possible.

After graduating and beginning a professional career, you might find it advisable to become actively involved in the planning of a new facility. If this should occur, and the analytical approach just discussed brings you and other users of the laboratory into close contact with the building design team, both groups will undoubtedly benefit. The ultimate goal will be the continuous refinement of the laboratory work space, in keeping with the exhortation of a famous laboratory scientist, Louis Pasteur:[4]

> I implore you, take some interest in those sacred dwellings meaningfully described as laboratories. Ask that they be multiplied and completed. They are the temples of the future, of riches and of comfort. There humanity grows greater, better, stronger; there she can learn to read the works of nature, works of progress and universal harmony, while humanity's own works are too often those of barbarism, of fanaticism and of destruction.

Figure 7.7 Computer Generated View of Laboratory (Courtesy CRSS)

Figure 7.8 Computer Generated View of Laboratory (Courtesy CRSS)

138

Notes and References

1. The sources you will find most helpful on the architecture of laboratory design are the following: "A Moving Target," *Architectural Record*, (February 1991), pp. 97–107; S. Braybrooke, *Design for Research: Principals of Laboratory Architecture*, New York: John Wiley & Sons, 1985; C. Collins and C. Gentry, eds., *Facilities Planning Handbook*, Tradeline, Inc., 1990; W. R. Ferguson, *Practical Laboratory Planning*, New York: John Wiley & Sons, 1973; R. Lees & A. F. Smith, eds., *Design, Construction and Refurbishment of Laboratories*, Chichester: Ellis Horwood Ltd., 1984; J. F. Munce, *Laboratory Planning*, London: Butterworths, 1962; "Inquiry: Laboratories," *Progressive Architecture*, (August 1990); T. Ruys, AIA, *Handbook of Facilities Planning, Volume 1, Laboratory Facilities*, New York: Van Nostrand Reinhold, 1991; *Planning Academic Research Facilities: A Guidebook* prepared for the National Science Foundation, Washington DC, March 1992; and "Laboratory Architecture," *Architecture*, (March 1993), pp. 123–127.

2. D. B. Brownlee and D. G. De Long, *Louis I. Kahn: In the Realm of Architecture*, New York: Rizzoli, 1991, p. 180.

3. The Nuffield Foundation, *The Design of Research Laboratories*, London: Oxford University Press, 1961, pp. 16–17.

4. The Nuffield Foundation, p. 7.

Two Assignments for Sophomores and Juniors: Laboratory Life

Assignment 1

Recall your history courses. There you might have studied the architecture of cathedrals, public buildings, and libraries. Do you recall any discussion of the architecture of laboratory buildings? Did you study any famous laboratory buildings by important architects? If not, why do you believe this important twentieth-century architectural building type is often overlooked? Work out your reasons with your laboratory group, and spend five minutes with your laboratory instructor discussing what you have theorized.

Assignment 2

In your university there are many science disciplines: physics, biology, chemistry, medicine, etc. Visit several of these departments and, in a notebook, write your observations about the differences and similarities in the design of the labs and offices. Talk to some of the people in the various departments so as to assess the merits of their laboratories. Make a five-minute report discussing your observations. Frame those issues that could make the labs you saw even better.

Three Assignments for Juniors and Seniors: Analyzing Laboratories

Assignment 1

Assess the various elements—fume hoods, electronic equipment, lab benches and tables, wall mounted cabinets and desk space—in the laboratory that you use. What is beneficial about the laboratory? What should, if funding could be found, be changed? Draw a diagram of some ideal relationships of these laboratory elements so that you can work most productively. Keep in mind the key issues discussed in this chapter. Write up your analysis using the two-page briefing memo discussed in Chapter 11. Be sure to attach your diagram.

Assignment 2

Interview your laboratory instructor on the best laboratory he or she has ever worked in. What made that laboratory so successful? Consider function, safety, and aesthetics as discussed in this Chapter. Use the interview to help you develop a kind of idealized laboratory or laboratory prototype. Make a five-minute report to your class on the results of the interview, using the presentation proposal format discussed in Chapter 11. In your proposal, illustrate the laboratory prototype that you developed.

Assignment 3

Recall the discussion of safety in Chapter 3. Assess the safety of your laboratory. Identify and diagram all lab components that contribute to the safety of the laboratory (fume hood, eyewash, safety shower, fire extinguisher, etc.). Identify and diagram all hazards of the laboratory (unvented solvents and chemical procedures, explosive materials, radioactive compounds, etc.). Explore with your group, in a five-minute report, all steps that have been taken in the design of your laboratory to promote safety. Identify ways to further improve the safety of your laboratory.

Design Challenge for Future Laboratories

Figure 1 below shows a 19th-century "general" laboratory, the prototype of the modern research laboratory. Unlike many labs today, the 19th-century lab was fairly static: walls were well-built and finishes permanent, the furnace with hood was built-in, there was little adjustability in the furniture, and storage space was fixed in place. The 19th-century lab reflected functional realities that would remain constant for the life of the building.

Figure 1. A 19th-Century General Laboratory
(Engraving from Alan E. Munby. Laboratories: Their Planning and Fittings. *London: G. Bell and Sons, LTD., 1921. xviii.)*

The development of the modern laboratory took a leap forward under the leadership of Thomas Edison, the inventing wizard. In New Jersey's Menlo Park, Edison set up an invention factory in 1876 that became, in time, a prototype of the large industrial research laboratory that would later be established by Kodak, General Electric, and Du Pont. As Paul S. Boyer writes, "Edison's laboratory demonstrated that the systematic use of science in support of industrial technology paid large dividends."[1]

(more)

Edison created an entire system, a facility that brought teams together could look not only at an idea but also at its application. To get to the solution of any problem as quickly as possible—this become the goal of research. The Manhattan Project accomplished this goal under J. Robert Oppenheimer; and NASA got us to the moon by the same agenda in the 1960s.

The Future Lab

Today, the growing probability of periodic lab rearrangement due to changes in project priorities has become an important consideration in lab design. The trend towards accommodating such shifts—whether technological, administrative, or economic—has put a higher premium on adaptability. The process of accommodating a range of laboratory needs within a fixed building shell has become increasingly important ever since the first modular labs attempted to integrate structure with adjustable furnishings and easily accessible mechanical services.

Today, however, the traditional idea of the laboratory module must be reevaluated. New types of instrumentation, robotics, and associated infrastructure demands; increasing concerns about life-cycle costs; uncertain and changing business priorities; the linkage of information systems enhancing computational power; the revolution occurring in scientific procedures themselves—all are forces that are causing architects and engineers to re-think the traditional view of the laboratory as a module.

The solution? New prototypes of lab space must be realized for the next generation of rapid change in science. The laboratory for the next generation of productive and creative science must be designed as a *platform*—a new working environment for science and engineering. The platform must incorporate systems—electronic, telecommunications, and piped pressure services—that enable multiple, new approaches to problems; the platforms must bring people together while isolating hazards and removing barriers.

It is the development of these platforms that constitutes the major design challenge for architects today.

1. Paul S. Boyer et al. *The Enduring vision: A History of the American People*. Lexington, MA: D. C. Heath, 1993. 599.

Making the Best of a Bad Situation: A Note on Laboratory Architecture from Bob Barat

As a researcher, it is clearly desirable to have your needs and opinions considered during the design of a new laboratory building. However, sometimes circumstance (just plain old bad timing) forces you to make do. I was faced with just this situation in my current laboratory building, The Advanced Technology Center (ATC).

The ATC was planned and constructed while I was off in graduate school. One of my colleagues, who was already an active researcher at the university I wished to join, was consulted during the design phase. During past years, he had to retrofit old labs to meet his needs. Now, he exercised this golden opportunity to order exactly what he wanted: low benches, high support racks, compressed air, cooling water —the works—all in one large room. When the ATC was completed, he moved in; in no time, he set up his equipment and restarted the experiments.

When I finally finished graduate school and became a faculty member with research plans, I looked to the ATC for laboratory space. Alas, all the good rooms were taken. All that was left was space which had originally been designed to house and support an experimental rotary kiln incinerator. The room for the kiln itself was very large, with a high head space. A small adjacent room was to serve as a gas cylinder storage room. It was designed to be explosion-proof, including a Halon fire extinguishing system based on ultraviolet flame detection, and a blow-out wall. The only utilities into this room were stub pipes sticking out of one wall. Even the electrical outlets had cumbersome explosion-proof receptacles which were only compatible with similarly designed expensive plugs.

I was offered this small room, which I gladly took. However, considering that my experimental combustion program called for an open flame burner probed by an ultraviolet laser beam, the Halon system was deactivated. Even though the room was small, I managed to set up two optical tables and the laser system. The experi-

(more)

ment is actually producing data, but not until much effort and time had been expended on making the room tolerable for my needs. Even dark clouds have silver linings. The one saving grace of the small room was its lack of windows. For the sensitive optical experiment I am performing, there is no problem with sunlight.

My anecdote is just one small instance of the ways in which chemical engineers might have to expend time and energy or be forced to curtail their endeavors in order to have the physical environment— the laboratory—accommodate their research. In the best of all possible worlds, however, the working environment should promote and facilitate work rather than impede it. This should be the primary objective of laboratory design.

8 Engineers and the Environment: Preventing Pollution at the Source

Daniel Watts
Executive Director
Air Emission Reduction Center

"The world is not so large as we intuitively believe—space can be as short as time."

Bill McKibben, *The End of Nature*

Preview

In Chapter 8, you'll have a chance to think about

- a framework for environmental decision making

- five environmental perspectives

- a case study of waste reduction activities

"Let's turn into yonder Arbor, for it is a clean and cool place."

Isaac Walton, *The Compleat Angler*

Part I: An Introduction to Pollution Prevention

Nature can speak with a voice as loud as an erupting volcano or a mighty wind storm. Nature can speak with a voice soft and insistent as a slowly migrating sand dune. No one needs to speak for nature; nature has a voice of its own.

In the same way, we all must speak to each other about our actions and the implications of our actions toward our continuing interaction with the natural world.

Today, people have recognized problems with that relationship. The challenges facing us include concerns about the quality of air and water. We are anxious about how to deal with the quantities of wastes we produce through our industrial activities and because of our private acts as consumers. Potential risks from contaminated sites remaining from past accidents or deliberate waste management practices continue to generate worry and fear.

None of these situations simply happened. Each resulted from a series of deliberate actions and decisions made by a sequence of people. The environmental problems of today, unexpected or unwanted as they may be, are the result of a collective of decisions made by people in the past.

One task for the scientist or engineer now is to understand the connection between the decisions of the past and the environmental problems of today. A greater task, perhaps, is to develop a decision making process that considers the potential environmental impacts of each possible course of action. In a larger sense, we all make environmental decisions. Every product purchase, every product use, and every product disposal represents an environmental decision. The role of the engineer, therefore, may be to anticipate the types of choices a consumer may make and work to reduce any negative environmental impact. [1]

A Framework for Environmental Decision Making

To be most effective, an environmental decision making should take place within an easily understood framework. The framework should reflect the procedures people and companies use in supplying goods and services and in selecting among them. The decisions should reflect the individual case of the current situation—whether, for example, to choose paper or plastic bags at the supermarket checkout or whether to choose toluene or ethyl acetate as a reaction solvent. The decisions should reflect also the more general case of long-range environmental implications.

Besides the immediate convenience of carrying home the purchase from the market, the more general cases include consideration of which choice uses less raw material, which uses less energy in production, which has secondary uses, which gives the best options at disposal. Additional decisions involve the immediate achievement of a better product yield: which solvent choice allows the reaction to be run with less energy input, which promotes recycling and reuse, which has less opportunity for fugitive emissions, which promotes easier equipment cleaning. In all cases, longer-range environmental considerations—beyond simply meeting immediate single situation needs—should play a role in decision making.

The issues frequently deemed important for engineers to consider in product, process, and plant design have included product manufacturing cost and product quality. The environmentally aware view suggests that environmental considerations of the product's manufacturing process, its use, and its eventual disposal should receive comparable weight. When compromises and trade-offs between objectives must be made, then environmental considerations should have the same value and importance as the other important parameters.

A common unifying goal for this type of environmental decision making may be to minimize the negative environmental impact of all human activities. Within the bounds of such a common goal, there are several different views of what should be achieved and of the context in which people should operate. Each has some different dimensions and characteristics. It is useful for an engineer to consider the implications of each of the perspectives as a base for personal decision making. [2]

Environmental Perspective 1: Sustainable Development

Sustainable development represents the view that the responses of society to meet the increasing needs and expectations of people through increased agriculture and manufacturing must be done in a balanced way. Specifically, there should be no net loss of natural resources. There should be no net degradation of environmental resources such as water and air. There should be a system for supplying human needs which is not limited by potential shortages of resources. There should be no wastes in an ideal situation; there should be a use for all products and by-products. Moreover, there should be continuing uses for all products. The concept of sustainable development is most often used to describe an environmental and developmental plan for a large area such as a country or a region. If sustainable development is to be possible for a large area, then each individual activity within that region must be carefully crafted into work toward that end. Fine tuning or redesign of each product or process within the region must be carried out if there is real hope that sustainable development can be realized.

Environmental Perspective 2: Clean Manufacturing

Clean manufacturing represents a concept of producing materials in an environmentally benign manner with regard to elimination or minimization of waste streams and emissions. Beyond that, clean manufacturing requires that the products not produce substantive negative environmental impact as a result of their use or disposal. Consideration of secondary uses and avenues for recovery of reusable components of the products should be a part of the clean manufacturing concept. In the same way, consideration of the environmental impacts of the processes used to manufacture or obtain the raw materials used in the product should be a factor in clean manufacturing. A concept related to clean manufacturing is environmentally conscious manufacturing.

Environmental Perspective 3: Pollution Prevention

Pollution prevention is a concept closely associated with an individual process or manufacturing operation. There has been substantial debate and discussion over the meaning of the term, and there are several views about what should be considered to be pollution prevention. One of the strictest definitions has been developed by the United States Environmental Protection Agency (EPA). Basically, the EPA definition considers pollution prevention to include changes in process or product which result in reduction *at the source* in the volume or degree of hazard of wastes and emissions. Recycling of materials in a closed loop within a process is also considered to be part of the concept of pollution prevention. However, other environmentally positive actions such as out-of-process recycling and material recovery and reuse are considered to be good options, but are not considered pollution prevention activities.

Environmental Perspective 4: Design for the Environment

Design for the environment is a concept which entails identifying specific environmental problems, identifying specific products or types of products which are responsible for the environmental problems, and designing replacement products which eliminate or reduce the environmental problem without contributing new problems in its place. For example, one problem which has been identified is that of mercury in the air emissions or in the ash from incineration of municipal solid waste. Analysis of the waste stream components which feed typical municipal waste incinerators reveals the presence of batteries. It is known that many common categories of batteries used by the consuming public contain mercury. A design for the environment approach in this case would consist of a redesign of the battery system to limit severely or to eliminate the mercury in the battery. Many manufacturers have already done this with resulting improvement in the mercury levels at incinerators. An alternative approach of developing a collection and reuse system for the components of batteries may qualify as a sustainable development or a clean manufacturing approach, but would not be considered an environmental design approach.

Environmental Perspective 5: Life Cycle Analysis

Another concept frequently discussed in the context of environmental decision making is life cycle analysis. This term refers to a systematic process to investigate and consider the environmental and energy impacts of the manufacture and use of products. This approach attempts to provide some quantitative comparisons of different strategies in product design and manufacture to provide a good basis for deciding on the most environmentally sound course of action. The approach is still relatively young and not enough data is available for easy and wide use of the technique. However, the potential and the intellectual approach provide a mechanism which all decision makers can apply.

These five different frameworks for environmental decision making range from choices having global impact to choices having immediate effect on a specific activity or operation. The engineer should be encouraged to develop a decision making philosophy which leads to sound choices for each specific activity and which leads to beneficial results on a global scale.

Moving from a World View to Discrete Local Action

The New Jersey Pollution Prevention Act of 1991 was created, in part, to transform the existing pollution control regulatory system into a pollution prevention system. The goal of the pollution prevention program is to encourage industry to study and change its operations to create smaller amounts of wastes. The Act also proposes a state-wide public policy goal of a 50% reduction over 5 years in the amount of hazardous wastes generated (as defined by the EPA's reporting regulations). The Act requires the preparation of pollution prevention plans by all companies that prepare and submit Toxic Release Inventory reports. A plan summary must be filed with the New Jersey Department of Environmental Protection and Energy. Annual progress reports must be prepared and filed.

For industries in Standard Industrial Classifications (SIC) 26, 28, 30, 33, and 34 (see Table 8.1), the pollution prevention plans must be completed and the plan summaries filed by July 1,

1994. The first five priority industries represent a diverse cross section of manufacturing in the state. Descriptions of the five major group SIC codes appear in Table 8.1.

Table 8.1 Descriptions of the Major Group SIC Codes of the Five Initial Priority Industries Under the New Jersey Pollution Prevention Act

SIC Major Group	Description
SIC 26	Paper and Allied Products
SIC 28	Chemicals and Allied Products
SIC 30	Rubber and Miscellaneous Plastics Products
SIC 33	Primary Metal Industries
SIC 34	Fabricated Metal Products, Except Machinery and Transportation Equipment

For industries in other classifications, the first due date is July 1, 1996. The writers of the Act designated the first group of industries for priority attention for two major reasons. Some make major contributions to the state's total releases. Others seem to use raw materials inefficiently based upon comparison of the amount of raw material with the output of product. These statistical assumptions suggest that this group of five industrial categories has the greatest immediate opportunities for pollution prevention. Therefore, the Act provides for their priority in implementation.

One of the jobs for an engineer working in any of these initial priority industries will be to reconcile the manufacturing objectives of the company with the requirements of the Pollution Prevention Act. All of the environmental decision making processes which seem desirable in abstract terms need to be tested in real situations.

Consideration of the five pollution prevention priority industries in New Jersey reveals that although the industries are significantly different in terms of the products they manufacture, they share a small number of production steps which are responsible for significant portions of the loading of releases and emissions to the environment. These generic processes certainly have specific applications to specific industries which must be examined individually. On the other hand, they also have common aspects which if examined from a pollution prevention perspective could have rapid and multi-industry impacts.

These generic cross-industry pollution prevention possibilities include the following:

- Low solvent or aqueous based coating operations

- Low solvent or aqueous based printing operations

- Low solvent or aqueous based gluing or adhesion processes

- Improved handling and transfer technology for volatile organic compounds

- Improved separations technology to return starting materials to processes and minimize their loss to emissions

- Improved volatile organics capture and recovery technology to facilitate the in-process return of volatiles during manufacturing operations

- Low solvent or aqueous based equipment cleaning technology

- Non-chlorinated solvent degreasing and surface cleaning technology

We turn now to a report and discussion of the type of evaluation for environmental decision making carried out at one manufacturing facility. The conclusions drawn and the recommendations made reflect the views and perspective of one group at one time. Another engineer with different perspectives may have different suggestions about improvements at this facility. The discussion illustrates application of some of the generic pollution prevention approaches described above. The assessment summarizes part of work done under Cooperative Agreement No. CR–815165 under the sponsorship of the New Jersey Department of Environmental Protection and Energy and the United States Environmental Protection Agency.

Part II: Pollution Prevention in Action—
A Case Study: Waste Reduction Activities and
Options for a Manufacturer of Paints

Facility Background

The facility in this case study is a producer of paints used primarily in the metal finishing industry. The paints are also used in automobile refinishing applications. This business requires production of a large variety of colors and finish types, most in relatively small quantities. The specifications of customers allow a very narrow range of variation in color and appearance of the finished product. This severely limits the flexibility the company has in changing production processes.

The facility is located in a rural/suburban area on a large tract of land. The facility was constructed originally for another type of manufacturing operation and has been retrofitted for the present paint production operation. The facility is clean and well-maintained, indicating that the management is aware of and concerned about any potential health, safety, or environmental implications of their operations. About 250 people are employed at this facility.

Manufacturing Processes

The production of the various types of paints is conceptually very simple. Required operations include mixing and blending, under carefully specified conditions, raw materials either purchased from vendors or shipped from other company sites. No manufacturing of paint constituents takes place at this facility. After formulation and blending, the paints are transferred to a variety of containers for shipment to the customer. The processing equipment is cleaned prior to preparation of the next batch. The cleaning operation typically includes multiple rinses with solvent in order to remove the pigments and additives remaining from the previous batch.

This manufacturing process can be seen as using a solvent or liquid carrier to dissolve or suspend the components of a coating system which will be deposited on a surface intended to be covered. This process is a large user of solvents. At present, the preponderance of the solvents used in these applications are organic. In the coatings industry there is a trend toward water-based products where customer demands and product performance make it possible. The technology for water-based coatings has not been sufficiently advanced to address all such demands and performance requirements. Therefore, solvent based paints and coatings will be required for some time.

Existing Waste Management Activities

The company has already instituted a program of pollution prevention. This is perhaps best illustrated by the acquisition and use of a large capacity still which allows recovery and reuse of the solvents from the equipment washing operations. Other pollution prevention efforts have been carried out in conjunction with the corporate research and development group. These efforts led to the reduction or elimination of the use of heavy metal-containing dyes and pigments in products produced by this facility.

The major waste from this facility that comes under Resource Conservation and Recovery Act (RCRA) jurisdiction is the still bottoms from the recovery/recycling/reuse of waste solvents from the equipment washing process. About 250 drums of this material are produced annually from the facility and are sent off-site for disposal. This quantity represents 10 to 20% of the volume of waste solvents which were sent for disposal prior to the installation of the distillation equipment.

Another waste stream results from quart-size quality control samples of finished batches which are retained at the facility for examination if customer problems or complaints come in about specific batches of paint. After the retention period, the samples are discarded as hazardous waste. The typical current practice is to recover the solvent from these retained samples through the solvent recovery system. There was no information available on the number of these samples obtained each year.

Another waste stream identified was a waste oil stream from equipment maintenance and repair. This stream averages 3 to 4 drums per year and is sent off-site for recycling and recovery.

The greatest pollution prevention challenge at this facility is not RCRA-type waste streams. Rather it consists of stack emissions and fugitive air emissions. Superfund Amendment and Reauthorization Act (SARA) Title III reporting and additional estimates indicate that approximately 200,000 lbs. of solvent are emitted to the atmosphere annually. The facility intends to address this situation using a pollution prevention approach.

Waste Minimization Opportunities

The type of waste currently generated by the facility, the source of the waste, the quantity of the waste, and the annual treatment and disposal costs are given in Table 8.2.

Table 8.2 Summary of Current Waste Generation

Waste Generated	Source of Waste	Annual Quantity Generated	Annual Waste Management Cost
Still Bottoms	Residue from solvent recovery and recycling	250 drums	$65,000
Waste Oil	Obtained from equipment maintenance and repair	4 drums	$130
Volatile Organic Chemicals	Fugitive and stack emissions of solvents used throughout the facility	200,000 lbs	$40,000 (This cost represents the estimated value of the solvents lost to the atmosphere.)

This particular facility presents a challenge in describing waste streams. The presence of an operating solvent recovery system means that the actual waste streams sent off site are relatively insignificant in terms of the total effluent from the process before the solvent distillation. Moreover, where there is a significant level of air emissions to be addressed, the meaning of the term waste management cost has to be strained to mean simply cost of the lost materials.

Table 8.3 shows the opportunities for waste minimization recommended for the facility. The type of waste, the minimization opportunity, the possible waste reduction and associated savings, the implementation cost, and the payback time are given in the table. The quantities of waste currently generated at the facility and possible waste reduction depend on the level of activity of the facility. All values should be considered in that context.

It should be noted that the economic savings of the minimization opportunity, in most cases, results from the need for less raw material and from reduced costs associated with waste treatment and disposal. It should be noted that the savings given for each opportunity reflect the savings achievable when implementing each waste minimization opportunity independently and do not reflect duplication of savings that would result when the opportunities are implemented in a package.

The cost savings are calculated both in terms of avoided costs of waste disposal and recovery of the value of raw material used again. Also, no equipment depreciation is factored into the calculations.

Table 8.3 Summary of Recommended Waste Minimization Opportunities

Waste Stream Reduction	Minimization Opportunity	Annual Waste Reduction Quantity	Annual Waste Reduction Percent	Net Annual Savings	Implementation Cost	Payback Years
Still Bottoms	Segregation and recovery of concentrations from washings of raw materials for reuse.	5 drums	2%	$3,100	$1,000	0.3
	Reprocessing of retained samples.	1 drum	0.5%	$323	$0	immediately
VOC Emissions	Develop and institute program of leak detection and correction. Reevaluate manufacturing processess in light of pollution prevention goal.	180,000 lbs	90%	$36,000	$150,000	4.2

The decision to add solvent distillation capabilities at the facility significantly reduced the volume of waste shipped from the site for treatment. It did, however, engender a new waste stream at the site. These still bottoms present a particular challenge from a waste reduction perspective: to minimize the still bottoms stream. Some possible options include identification of beneficial uses for the still bottoms, recovery of valuable materials from the bottoms, and change in operating practices to reduce the quantity and type of materials which appear in the still bottoms.

In the absence of specific information about the content of the still bottoms (which would be variable at best), it is not possible to suggest specific options regarding beneficial uses for this material. Similarly, it is not possible to discuss specific options for recovery of valuable materials from the bottoms. These questions can be effectively addressed after data about the composition of the still bottoms is collected. In general, it is assumed that the materials in the still bottoms consist of residues which are contained in the solvents as they enter the facility, product residues from equipment rinsing and cleaning, and manufacturing residues from disposal of products or raw materials. Modification of equipment rinsing and cleaning practices to reduce the amount of solids in the rinses would result in a decrease in the quantity of solids in the still bottoms.

One approach which may accomplish some of the objective to minimize the still bottoms stream would be to segregate washings from equipment used for transfer of raw materials from washings of equipment used for finished batches. Such segregation and distillation of raw material solutions should result in concentrated solutions of the raw materials which could be used in production, rather than disposed.

The majority of the still bottoms result from washing of equipment from the finished batches of coatings. The best opportunity for reduction of this component lies in scheduling of batches in terms of colors and coating types. If batches were appropriately scheduled, less rinsing may be required, particularly in moving from lighter colors to darker colors. On the other hand, it must be recognized that this facility has very tight specifications for color reproducibility because many of their customers do color matching. From a total pollution prevention perspective, it may be preferable to rinse the equipment thoroughly and collect the resulting still bottoms rather than risk the potential disposal of an entire batch of paint.

For impact on emission reduction, attention should be given to the sources of the emissions and on the potential options for emission reduction. As indicated previously, more than 200,000 lbs of SARA 313 emissions are reported annually from this facility. Approximately 70% of this total amount represents fugitive air emissions. The material emitted in largest quantity is acetone, representing about 50% of the fugitive air emissions and about 48% of the total emissions from the facility.

The Company's Approach: Directions

Before developing a slate of options for addressing this challenge, the company examined the question of how such large losses could have been accepted for so many years. All of the manufacturing operations met the company standards for material use; therefore, there was no reason to question the quantity of materials purchased and the quantity which actually went into the product.

Upon further questioning about how the company manufacturing standards were determined, the company concluded that when the product was first manufactured, careful records were kept and maintained for the first three or four batches. These records included information about materials used. Then with the addition of a slight margin for error, these quantities became the manufacturing standard. This reveals that procedures used in the past—which may contribute, as in this case, to elevated levels of fugitive emissions—are often perpetuated unless new questions are raised by an "outside" review process such as this one.

In advancing the list of options, the company reviewed the list of volatile chemicals which make up the fugitive air emissions. The company found that the list consisted of the solvents which are used in the manufacture of the coating products at the facility. In addition, acetone is used as a solvent for equipment cleaning. This kind of solvent use provides two different avenues to be explored for pollution prevention options. The solvent used in direct production is essentially fixed in terms of how much solvent must be in the product shipped to the customers; therefore, any reduction in emissions from this part of the operation must result from changes in losses from spills and leaks, incomplete transfers, and evaporation.

Because much of the material flow in the facility is a mechanized movement from large storage tanks to production vessels, there are opportunities for leaks at seals and connections. A high priority option could be to check the entire solvent supply system for leaks. Based upon experiences at other facilities, particular attention should be given to seals and to pumps. Regular inspection for such leaks should be a part of the program.

There are also opportunities for evaporative losses when the production vessels are being filled and operated. Certainly, air displacement is necessary in the tanks to allow proper filling. Both capping the tanks and a vapor recovery system utilizing a condenser could have a significant impact upon evaporative losses. Depending upon the quality of such condensate, it could be returned directly to the production vessel resulting in an immediate reduction in total material used for each batch. Alternatively, the recovered solvent could be sent to the distillation process for purification prior to reuse.

Evaporative losses can also occur when the containers which go to the consumers are filled. A contained process for this filling operation which would allow for solvent recovery by condensation could be investigated. Alternatively, smaller container openings would reduce evaporative surfaces and, consequently, fugitive air emissions. Process engineering and design efforts will be required in order to address these options.

Conclusions to the Case Study

Three broad conclusions about the nature of pollution prevention can be drawn from this case study.

First, it is clear that pollution prevention is a complex process requiring students of chemical engineering to think critically. The notion of preventing pollution appears, at first, simplistic. However, upon further reflection, the facts reveal that those involved must make a complex set of decisions that are, at once, economic, social, and technical. In this case study, for example, the ability to understand the nature of reducing a waste stream is a technical issue tied to the economics of this reduction. Students must therefore have a knowledge of the technical side of their field as well as the ways that their field operates in societal contexts.

Second, it is clear that collaborative decision making of the kind described in Chapter 3 is essential to the process of pollution prevention. Gone is the distinction between the university researcher, the field engineer, and the owner of a company. All must work collaboratively if complex solutions to complex problems are to be found.

Third, it is clear that pollution prevention is an idea whose time has come. The costs of remediation of waste have risen sharply; and the effects of environmental pollution—on nature itself, on the ways that we live in nature—are already staggering. A new age has begun, and it is up to students such as you to lead this new environmental effort.

Notes and References

1. Three books that will be of interest for students of the environment will be the following: Bill McKibben, *The End of Nature*, New York: Anchor, 1989; Lester R. Brown, Christopher Flaven, and Sandra Postel, *Saving the Planet: How to Shape an Environmentally Sustainable Global Economy*, New York: Norton, 1991; and Donella H. Meadows, *The Global Citizen*, Washington, DC: Island Press, 1991. Of interest also are the annual additions of *Environment*. The 11th edition (1992–1993), edited by John L. Allen, is available through the Dushkin Publishing Group.

2. For general guides to the concept of pollution prevention, see the following: Michael G. Royston, *Pollution Prevention Pays*, New York: Pergamon Press, 1979; U.S. Congress, Office of Technology Assessment, *Serious Reduction of Hazardous Waste for Pollution Prevention and Industrial Efficiency*, OTA-ITE-317, Washington, D.C.: U.S. Government Printing Office, September 1986; and U.S. Environmental Protection Agency Office of Pollution Prevention, *Pollution Prevention 1991: Progress on Reducing Industrial Pollutants*, Washington, D.C.: U.S. Government Printing Office, October 1991.

Assignment for Sophomores, Juniors, and Seniors
Newark Radiator Repair

This case study should be discussed over two class periods. A draft of the proposal should be submitted and peer reviewed in the third class period. The final proposal should be submitted at the beginning of the fourth class period.

The Context

You are a graduating senior at New Jersey Institute of Technology. You have worked your way through college, and you are presently employed as a shop manager at Newark Radiator Repair. This small business specializes in the repair of automobile and truck radiators. These radiators are made of steel and copper and are held together with a tin/lead solder. On a number of occasions, the owner has expressed to you a commitment to the environment. As an environmental advocate, you want to help your boss express his commitment by reducing the hazardous waste generated by the process of radiator repair.

The Overall Task

In this case study, you will write a proposal to the owner of Newark Radiator Repair. You will propose that the shop re-design its repair process.

A Procedure for Approaching the Newark Radiator Repair Case Study

Phase 1. Identifying the Repair Process

When a radiator comes into the shop, it first has to be drained of anti-freeze. The tank then has to be boiled out, tested, and repaired. Paint then is sprayed on the repaired radiator.

Local, state, and federal law, however, prohibits the discharge of antifreeze. When the tank is boiled-out and tested, the scouring process removes rust, oil, and dirt. However, the boiling also removes zinc, cadmium, iron, and lead.

(more)

This wastewater discharge cannot be simply sewered. When the tank is re-paired, a lead-based solder is used. When the radiator is painted after repair, a solvent-based paint is used.

Newark Radiator Repair is fully regulated under the Resource Conservation and Recovery Act (RCRA) as a hazardous waste generator. The shop dis-charges the anti-freeze and the wastewater from the scouring to a public treat-ment facility.

Task 1—Define the Repair Process. Your first job in preparing the proposal is to *draw a flowchart* of the repair of a radiator. In the flow-chart, be sure to identify each step of the repair process. Also, illustrate what presently happens to the discharge from the shop.

Phase 2. Identifying the Owner's Motivation

If the shop is presently operating under the RCRA, then why should the owner investigate any pollution prevention measures at all?

Task 2—Define the Motivation. *In a well-developed paragraph*, specu-late on the reasons for the owner's interest in designing a new process. Be sure to consider these motivations: legal compliance, environmen-tally-conscious advertising as a way to increase business, love of technological innovation, and humanistic devotion to the welfare of others.

Phase 3. Identifying Alternatives

Because the re-design process is complex, you decide to contact the New Jersey Technical Assistance Program for Pollution Prevention. After studying the repair process at the shop, a representative from the Program suggests a number of alternatives that would prevent the creation of hazardous waste.

First, the representative suggests that the owner purchase an anti-freeze recy-cling unit. Second, the representative suggests that the radiators be cleaned manually instead of by the chemically-driven scouring process that produces the zinc, iron, cadmium, and lead. Third, the representative suggests that a lead-free solder be used. Fourth, the representative suggested that a latex-based paint be used.

(more)

Assignment for Sophomores, Juniors, and Seniors
Newark Radiator Repair

Task 3—Identify the Alternative Process. In Task 1 you drew a flow-chart of the repair of a radiator. Now, *draw a second flowchart*, that identifies the pollution prevention alternatives to each step of the repair process.

Phase 4. Identifying Possible Problems

Before your boss initiates any of these new processes, you are going to have to discuss the possible problems that might arise with the new methods. This strategy is simply good business. When potential problems are realistically anticipated, a pollution prevention venture has a greater change of success.

Task 4—Define Possible Problems. *In a well-developed paragraph*, speculate on the kinds of problems that might arise from the suggestions of the representative of the New Jersey Technical Assistance Program for Pollution Prevention. Specifically, you should consider variables such as labor, cost, and technical reliability.

Phase 5. Writing the Recommendation

To write the proposal, we suggest you begin with an introduction in which you provide an overview of your ideas. Next, you might isolate the rewards the owner might gain as a result of modifying the radiator repair process. Third, you should identify any problems that may be encountered in modifying the repair process. You might end with a set of steps you have formulated so that the owner will know what procedures should be taken to modify the repair process.

Task 5—Write the Proposal. In no more than five single-spaced pages, *compose a proposal* to the owner of Newark Radiator Repair. Be sure to include and explain the two flowcharts you have drawn. For help with writing the proposal, refer to Chapter 11.

(more)

Assignment for Sophomores, Juniors, and Seniors
Newark Radiator Repair

Phase 6. Editing the Recommendation

After drafting the proposal, you should have a classmate review your work. Specifically, your classmate should address the following questions:

1. Will the owner of the shop be favorably disposed toward the proposal in the five seconds after he begins reading it? Why or why not?

2. Are there clearly delineated sections of the proposal that are identified by headings? Are these sections clearly written and topically unified?

3. Have the figures been integrated into the text and fully explained?

4. Is there a clear statement and discussion of gains that the owner might realize from the new repair process?

5. Is there an honest statement of possible problems the owner might face in implementing the new process?

6. Is there a statement of the reasons that the owner should nevertheless begin the new pollution prevention process?

7. After studying the proposal, will the owner feel that the writer is competent enough to begin the pollution prevention venture?

Task 6—Peer Review the Proposal. *Design between six to eight questions* that address specific areas you would like to have reviewed in your own paper. When you submit your paper for peer review, submit these questions to your reader and ask for a detailed response.

Phase 7. Submitting the Proposal to Your Instructor

On the assigned date, *submit the revised, final proposal* to your instructor.

Task 7—Debriefing. Now that you have submitted the final proposal for a grade, *spend some time in class discussing* what you have learned in the case study. Specifically, you should discuss the ways that small businesses are unique in addressing pollution prevention issues. You should also infer a set of characteristics that describe the nature of small business pollution prevention initiatives.

(more)

Assignment for Sophomores, Juniors, and Seniors
Newark Radiator Repair

Task 8—Imagining the Future. The owner of Newark Radiator Repair has done what he can, but he is limited in his ability to reduce waste because of the hazardous materials that enter the shop when the radiator is delivered through the door. *Spend some time discussing* what manufacturing alternatives might exist in the future for the manufacture of radiators? Because pollution prevention is a series of complex trade-offs, what new problems might a manufacturer face in the design of a new radiator?

(Background material for this case study may be found in the following article: Kevin F. Gashlin and Daniel J. Watts, "Modifying Process & Product: Two Pollution Prevention Case Studies." *Journal of Environmental Regulation* [Winter 1992/193]: 139–149. This exercise was developed under grant #992903 from the Environmental Protection Agency for Science, Technology, and Society Curriculum Transformation.)

Databases for Pollution Prevention

How do we study the gains made by pollution prevention? While specific studies can be made in individual cases, national trends are more difficult to study.

One approach to large-scale study is to identify databases that deal with pollution prevention. In 1991, the United States Environmental Protection Agency (EPA) took just that approach. The EPA identified five databases:

the Toxics Release Inventory (covering 320 chemicals from 21,000 manufacturing facilities, reporting from 1987 to the present)

the Hazardous Waste Generator Survey (a one-time survey of a sample of 16,572 facilities generating volumes of hazardous solid and liquid wastes; the data covered are for the calendar years 1985 and 1986)

the Hazardous Waste Biennial Report (covering the same domain and time period of reporting as the generator survey)

the Chemical Manufacturer's Survey (covering the annual volume of solid and liquid waste from approximately 600 member chemical plants, reporting from 1981 to the present), and

the American Petroleum Institute's Survey (covering the annual volume of solid and liquid wastes from 176 oil refineries, reporting from 1987 to the present).

These databases cover such information as throughput data, source reduction, waste management, and facility releases and transfers.

The EPA study proved interesting in that it revealed just how difficult it is systematically to collect large-scale information on pollution prevention. For example, the Toxics Release Inventory indicated a decline in reported release quantities. Nevertheless, there was insufficient information to say exactly what was being decreased because of lack of methodological reporting.

(more)

The EPA report tells us that there are encouraging signs that pollution prevention is occurring and that the future holds promise especially for the manufacturing sector. In addition, the Federal Pollution Prevention Act will provide further information in that pollution prevention reporting will become mandatory. (So, take another writing class because professional writers will be needed for those reports.)

It is likely, the EPA report concludes, that data will shortly be available "to allow both the assessment of the extent of industrial pollution progress nationally for the manufacturing sector, as well as at individual facilities in the sector."

Source: U.S. Environmental Protection Agency Office of Pollution Prevention. *Pollution Prevention 1991: Progress on Reducing Industrial Pollutants.* Washington, D.C.: U.S. Government Printing Office, October 1991. 15-35.

9 Communicating Information in Chemical Engineering

"To be only functionally literate in a hyper-literate society is to live as an oppressed stranger in an overwhelming world."

Andrea A. Lunsford, Helen Moglen, James Slevin,
The Right to Literacy

Preview

In Chapter 9, you'll have a chance to think about

- communication as the vehicle for critical thinking,

- the importance of the ability to write clearly and speak forcefully

- the relationship between writing and research

- definitions of literacy and technical writing

- a communications model for chemical engineers

- analyzing the needs of the reader

- the role the writer plays

- the forces that act on the reader/writer relationship

- a writing process for chemical engineering students

"A Preacher that was to preach to procure the approbation of a Parish, that he might be their Lecturer, had got from his Fellow-pupil the copy of a Sermon that was first preached with a great commendation by him that composed and preach'd it; and though the borrower of it preach'd it word for word, as it was at first, yet it was utterly disliked as it was preached by the second: which the sermon-borrower complained of to the lender of it, and was thus answered; I lent you indeed my Fiddle, but not my Fiddlestick; for you are to know, that every one cannot make musick with my words, which are fitted for my own mouth.

Isaac Walton, *The Compleat Angler*

Part I: Introduction

How is Research Possible? The Significance of Writing

The characteristics of our modern scientific culture are difficult to describe. To trace the rise of science is even more challenging. You have had, no doubt, experience in explaining the emergence of the scientific method. You have probably learned to discuss the importance of Copernicus' publication in 1543 of *On the Revolutions of the Heavenly Spheres*, the significance of Bacon's use of the scientific method in the early 17th century, and the contribution of the national scientific societies that flourished throughout Europe in the 19th century. You can probably account for the influence of the 19th century industrial revolution on science. And, along with Robert Oppenheimer, you can warn about the dangers of technology that emerged after the development of the atomic bomb. Perhaps you can even use Stephen Smale's horse shoe to explain chaos theory.[1]

While the traditional academic accounts for the rise of scientific thought are helpful, they often overlook the obvious. In order to address this oversight, Bruno Latour, a sociologist, stresses that we must take writing into account when we discuss the development of the modern scientific enterprise of which you are part.[2]

Latour argues that we need "to seek more mundane explanations" when understanding the chemical engineering paradigm of Chapter 3.[3] Latour emphasizes the following as contributing to the kind of rigorous research we associate with chemical engineering:

> *Inscriptions.* In the modern research effort, everything must be turned into writing. Rats and chemicals alike, Latour reminds us, are eventually transformed into paper. Researchers hold dear the extraction of precious pieces of information that allow hard facts. Thus, when you study the principles of mass transfer as evinced in a Karr Column, you transform your findings into print by means of the reporting structures discussed in Chapter 11. The print becomes the reality.

> *Print.* Print allows you to mobilize your ideas. As you create ideas by inscription (writing), you distribute them by means of print. Indeed, it may be helpful for you to think about the kind of print culture of a chemical engineering laboratory as it is created by laboratory notebooks and research reports. Why is a print culture important? You may have ideas about thermodynamic principles, but without encoding your ideas on the page, that knowledge stays local and temporary.

It becomes clear that writing is, at least in part, crucial to the creation of meaning that is created in the research process.

How are Facts Constructed? Beyond Elementalism, Toward Technical Writing

If we turn to an earlier work by Latour, *Laboratory Life*, we find another provocative argument for the necessity of writing in research.[4] After working for two years in a laboratory of the Salk Institute, Latour and his colleague, Steve Woolgar, concluded that "scientific activity is not 'about nature,' it is a fierce fight to *construct* reality. The *laboratory* is the workplace and the set of productive forces, which makes construction possible."[5] The order that we associate with valid engineering practice, then, becomes a created order, made possible by the power of writing.

Acts of communication are far more than exercises in correctness (e.g. where to place the comma, how to spell, when to capitalize). Writing, in contrast, is the very vehicle that makes *order* possible. It therefore follows that, unless you are able to write effectively, you will never be able to participate fully in the profession of chemical engineering.

You need to move beyond a mere mechanical definition of writing to writing as a means to create order. In order for you to do this, we need to define writing itself. We will do this by means of four principles that will help you to move beyond misleading notions of writing that you may carry around with you:

1. *Writing is a complex craft that you can learn.* Developed by the Sumerians in 3200 B.C., writing emerges with the beginning of civilization itself. First used to keep track of trade, writing developed as a means for both commerce and entertainment. Today, the act of writing is so common that we almost overlook its importance as a powerful technology. To utilize this technology, you must rid yourself of the notion that writing is a gift distributed at birth to some (English teachers) and denied to others (chemical engineers). Writing, while a powerful tool, is none the less an activity created by craft. Regardless of your experiences with that subject called English, you can learn to be an effective writer.

2. *Writing shapes your thinking.* You have, no doubt, had the experience of writing down your ideas about an experiment you have conducted, reading those ideas, and realizing that the words have failed you. As you began to revise—to get the message right—you noted that your process of writing, in fact, caused you to write more precisely. Often, only when you have completed the laboratory paper do you feel that you have truly understood the problem at hand. If this process is familiar to you, it is because you have experienced a great benefit of writing: it helps you think more carefully. As literacy theorist Walter Ong observes, "writing restructures consciousness."[6]

3. *Acceptability in writing is determined by your audience.* Different audiences value different kinds of writing. A chemical engineer, for example, may work on a lead in a briefing memo that can be read in five seconds, while a writer working on an article for the *New Yorker* magazine may write an essay with a three-paragraph introduction. Because you have spent so much time in traditional English composition classrooms, you will probably find the kinds of reporting structures described in Chapter 11 beyond your experience. Nevertheless, you need to broaden your repertoire beyond the essay, the term paper, and the lab report if you are to survive in your own community of chemical engineering.[7]

4. *Writing empowers you.* If you cannot write effectively, you will probably not progress in your profession. If you write in order to learn, you will more fully understand the complexities of your field. If you write in order to convey information, you will be in control of your work environment. An informative memo will create an environment in which your efforts will be more easily understood; you will begin to shape your organization through the use of language. Rather than being controlled, you will be in control.[8]

As you abandon the simplistic, mechanical view of writing, you must also develop an appropriate definition of writing. This is best done if you understand that there are multiple levels of literacy.

Levels of Literacy

The figure below illustrates the multiple literacies you have—and will—experience in your career as a chemical engineer:[9]

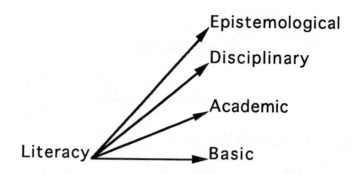

Figure 9.1 Levels of Literacy

Because you have achieved sophomore, junior, or senior standing in your chemical engineering, you have achieved mastery of *basic* literacy. That is, you can produce a grammatically correct, fluent, organized, autobiographical essay under controlled conditions. Because you have been in the university for at least a year, you have also had some experience with *academic* literacy. To further help you develop this level of literacy, we have analyzed and presented various academic forms of writing in Chapter 10.

The level of literacy you are working with in this text may be called *disciplinary* literacy. This kind of writing requires that you be familiar with the various aims of writing and the reporting structures used in chemical engineering. Chapter 11 is devoted to a discussion of these structures.

The highest level of literacy we term *epistemological*. That is, we believe that literacy is at its highest level when you gain an understanding about yourself by means of writing about your journey through chemical engineering. This point is worth an illustration.

Epistemological literacy can be seen in Primo Levi's book *The Periodic Table*. A chemist by profession, Levi ends his work with a piece entitled "Carbon".[10] He there reveals his purpose in writing:

> The reader, at this point, will have realized for some time now that this is not a chemical treatise: my presumption does not reach so far . . . . Nor is it an autobiography, save in the partial and symbolic limits in which every piece of writing is autobiographical, indeed every human work; but it is in some fashion a history.

> It is—or would have liked to be—micro-history, the history of a trade and its defeats, victories, and miseries, such as everyone wants to tell when he feels close to concluding the arc of his career, and art ceases to be long. Having reached this point in life, what chemist, facing the Periodic Table, or the monumental indices of Beilstein or Landolt, does not perceive scattered among them the sad tatters, or trophies, of his own professional past. He only has to leaf through any treatise and memories rise up in bunches: there is among us he who has tied his destiny, indelibly, to bromine or to propylene, or the -NCE group, or

glutamic acid; and every chemistry student, faced by almost any treatise, should be aware that on one of those pages, perhaps in a single line, formula, or word, his future is written in indecipherable characters, which, however, will become clear 'afterward'. . . .

> So it happens, therefore, that every element says something to someone (something different to each) like the mountain valleys or beaches visited in youth.[11]

As he achieved knowledge through his profession, Levi was able to achieve knowledge of himself. Implicit in our definition of literacy, therefore, is our hope that your training in chemical engineering will be used, ultimately, to help you better understand your own place in the world.

With these levels of literacy defined, we can now turn to the level of literacy most important to you at this stage of your career: the disciplinary literacy implicit in technical writing.

Technical Writing: History and Definition

Writing on technical subjects began in antiquity. Early documents include *De Architectura* by Vitrivius (c. 50–26 B.C.) and *De aquae ductu* by Frontius (c. A.D. 40–103). In the medieval period Chaucer produced his *Treatise on the Astrolabe* (1391). In the seventeenth century, the rise of modern science was accompanied by Francis Bacon's observations on style in *The Advancement of Learning* (1605). The origin of the Royal Society, chartered in 1662, had an influence on scientific writing. Joseph Priestly deserves attention for his uses of argument in the eighteenth century. In the nineteenth and twentieth centuries the research and interpretative works focusing on technology become too numerous to cite.[12]

The emergence of technical writing as a profession, however, is distinctly modern, and can be placed at the end of World War II. Robert J. Connors writes of the period that:

> During six years, necessity had mothered thousands of frightful and complex machines, and the need for technical communications had never been greater. (In 1939, officers were ordered to prepare for the war by sharpening their swords—an eloquent example of how much technology had changed the world by the time of Nagasaki six years later.) Technical writers were in great demand during the war, for each new airplane, gun, bomb, and machine needed a manual written for it, and the centrality of the lucid explicator was obvious as never before.[13]

As further evidence of the growth of technology during the 1940s, Edwin T. Layton finds that the post-World War II period was one of tremendous expansion of the engineering profession. In 1940 there were 260,000 engineers in America; by 1950, that number passed the half-million mark. By 1960 there were over 800,000 American engineers. The rapid growth of the profession produced a need for those who could translate technical information for a variety of audiences. As General Electric, Westinghouse, and General Motors opened separate departments of technical writing, the modern profession of technical writing emerged.[14]

Interest in technical writing instruction may be said to have begun at the turn of the century with the work of Samuel Earle of Tufts College. In response to the economic prosperity and technological growth at mid-century, technical writing instruction truly began to flourish in the 1950s. In 1950 the Society for Technical Writers was formed, and in that same year Rensselaer Polytechnic Institute began the first masters degree program in technical writing. By 1957, nearly all colleges offered a technical writing course, and 64% of all engineering schools made that course a requirement during the junior and senior year. In 1973 the first professional organization of technical writing instructors was formed, the Association of Teachers of Technical Writing.

And in 1976 even the Modern Language Association—long the stronghold of the ultra-conservative profession of English—recognized the existence of technical writing when the first panel concerning issues in technical writing was offered at its national convention.[15]

How is technical writing viewed today? Often, it is taught as a separate course, completely separate from students' disciplines.

It is pursued as an exercise in itself in which models and forms of nondisciplinary writing are used. For example, instead of presenting your experimental results in the form of an engineering user's manual within the context of your laboratory course, you might be asked to describe a potato peeler or a paper clip according to some model. In other words, the audience—and the content—are rarely considered.

We are against such approaches. We believe that the purpose of writing and the form that it takes must be inseparable. Technical writing is most effectively taught within the context of your discipline, within your particular engineering course. For example, in Chapter 11 you will be asked to write up the results of your experiment as a request for funding (proposal). You will have to consider both the argumentative purpose for the writing as well as the form of report. As you will see, when purpose and form are merged, technical writing instruction becomes real, exciting, and meaningful to *you*.

Below are two guidelines by which you should approach technical communications:

Guideline 1. *Abandon a mechanistic approach to writing*. Think back and you'll hear the voices: "Two comma splices, three misspelled words, one fragment—that adds up to an F." Or: "A paragraph always has five sentences." You remember the red marks, the obsessively neurotic focus on the grammar, the absence of positive comment on your content. No wonder so many of you express so many negative feelings about writing.

Therefore, it is crucial that you view writing as the complex, significant communicative act that it is. When you write up your experiments according to the forms we've provided in Chapter 11, realize that your task is to present significant ideas in a clear and compelling way. Plan, draft, review, and revise your work according to the process we suggest in Part II of this chapter. Now, this is not a licence to forget all about correct spelling, grammar, and so forth. What we are saying is that, when you realize the significance of what you are doing, correctness will naturally follow.

Guideline 2. *Reflect on the significance of writing and communication*. Earlier in this chapter, we presented four reasons for the significance of writing. First, we proposed that writing, though complex, is a craft that can be learned. Second, writing shapes your thinking by causing you to form hypotheses, test them, refine them, and communicate the resulting knowledge to others. Third, writing takes place in communities. Your writing will be ineffective if you ignore the needs of your audience. Fourth, writing will empower you, giving your ideas recognition, and you influence. It should be clear that the true significance of writing rests in its power to transform both you and your audience.

Our point? Language is your most powerful tool. No matter how poor your experiences with writing and speaking have been, you must put them behind you. Make use of the approaches to critical thinking and communication described in this book, and you will find that you are in control of a technology as powerful as any you will encounter in the laboratory.

With this background in mind, we now offer a definition of technical writing in the laboratory setting:

> Technical writing in the chemical engineering laboratory is that kind of writing which effectively translates your work and results for the technical, organizational, and cultural communities (audiences) which will read about and be influenced by your laboratory research.[16]

In order to demonstrate this definition, we need now to turn to its placement within a communications model.

A Communications Model for Technical Writing

You (the writer) and your audiences (the readers) strive toward an understanding, as shown in Figure 9.2. This relationship is affected by three cultures, or forces.

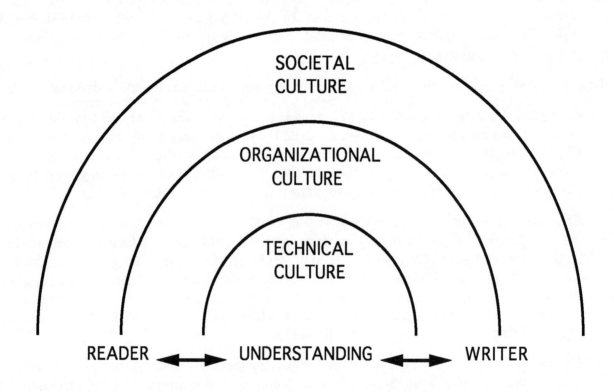

Figure 9.2 A Communications Model for Technical Writing

The essence of the relationship between the reader and writer is *understanding*. As the model illustrates, there are three forces affecting the level of understanding.

> The outermost force is that of *societal culture*. Societal culture is that vast arena in which all technology and organization exist. When, for example, Robert M. White, former administrator of the National Oceanic and Atmospheric Administration, reminds us that our national concerns about industrial competitiveness are linked to a global picture, he is noting societal culture. Try as we might, there is simply nowhere to hide from this force. In addition, we must acknowledge that the U.S. era of postwar prosperity is over: at the

end of World War II, the U.S. accounted for over 40% of the world's gross national product; by 1980 the U.S. share had fallen to 20%. To bridge the gap between national desires and global realities, chemical engineering is exactly the kind of activity, White believes, that will be called upon.[17]

The second force in the model is that exerted by *organizational culture*. Here we have the force exerted by the kind of modern corporate culture that prizes the brand of excellence that Peters and Waterman identified in their best-seller, *In Search of Excellence*. Defined by Siegfried Strufert and Robert W. Swezey as cognitive complexity, the attributes most prized by the organization include creativity, the use of strategy, and leadership.[18]

The third force is that exerted by the *technical culture* of chemical engineering itself with its methods, constraints, and agendas. According to the legendary author Isaac Asimov, the future for chemical engineering will hold research in fusion and an increased demand for sophisticated bioengineering. So, while the principles of chemical engineering remain the same, the research agenda might, in time, shift.[19]

These three forces exert great pressure on the understanding between the reader and writer. In addition, there are three elements to notice about this relationship.

First, *it is an interactive relationship*. The reader brings certain expectations and abilities to a document. Perhaps the reader is part of the technical culture and might then be concerned only with the formal aspects of your laboratory work such as the construction of the apparatus. If a member of the organizational culture, your reader might have an M.B.A. and be interested more in cost than in equation derivations. A reader who is a member of the broader societal culture may be interested in the biodegradability of your product or its international marketability. Needless to say, these forces pose enormous difficulties for the writer, difficulties we have tried to anticipate by means of the reporting structures provided for you in chapters 10 and 11.

Second, *the relationship is not expressed hierarchically*. That is, the writer is neither in a superior position to the reader—explaining technology to the naive citizens at the town meeting—or in an inferior relationship—flagellating himself on matters of cost before an organizational manager. Rather, the relationship is one of equality in which the primary interest is that of communicating information.

Third, *the relationship assumes that meaning emerges in a transaction between readers and writers*. Often, students assume that the page itself is the vehicle for transmitting information. If it is on the page, students seem to feel, then their responsibility is at an end. In our model, we stress that your responsibility as a writer is not ended until information is conveyed. This means that the reader is powerful and must be taken into account if the communicative act is to be successful and if understanding is to be achieved.

Part II: Practical Guidelines

Applications of Communications Theory

In Figure 9.2 we explained the theoretical communications model in terms of three components: (1) the role of the reader, (2) the role of the writer, and (3) the forces that act on the level of understanding between the reader and writer. We now want to examine further the practical implications of Figure 9.2.

Yet why, you might ask, do we not get right to the point and discuss the reporting structures provided in Chapters 10 and 11? Unless you understand the contingencies upon which writing is based, you are likely to interpret the kinds of reports as formulae. They are not. The documents we supply are *a* set of samples, not *the* definitive set. Yes, they are good samples and, yes, they represent the kinds of documents you will find in the field of chemical engineering.

Nevertheless, the interactive relationship of reader, writer, and context varies across time and place. A briefing memo might very well take a different form at Merck than it does at Exxon. If you understand that the forms may vary but the essences will not, you will approach the reporting structures with an inquiring attitude, that sense of critical thought that is at the core of this book. Therefore, the following two sections will help you to build and strengthen the understanding between reader and writer by discussing the needs of your audience (the reader) and your obligations as a writer.

The Needs of the Reader: A Question of Audience

The first issue you must address in writing is the identification of your audience. In 1976 John C. Mathes and Dwight W. Stevenson pioneered a systematic method of audience analysis. Their elegant three-step procedure remains unsurpassed in its clarity and richness; therefore, we want to present their system here so that you can develop a planned method of audience analysis. Mathes and Stevenson's method also has an additional bonus: the researchers used a chemical engineer's communications situation to explain their technique.[20]

Step 1. Prepare an organization chart with you in the center. The first step requires you to identify yourself relative to those who would read your report. Your first impulse might be to construct an organizational chart such as the one shown in Figure 9.3. Mathes and Stevenson warn us that such a conventional chart is not helpful, and might actually be misleading.

First, the organizational chart identifies complex bureaucratic units rather than the *real people* who will read your report. Second, you, the writer, are at the bottom, inferior to all potential readers. In reality, your readers are not necessarily superior, inferior, or equal to you; rather, they are either near or distant from the kind of information you will present.

As an alternative, Mathes and Stevenson propose that you use the egocentric organizational chart given in Figure 9.4. Here, you start at the center.

You must think of four degrees of distance between you and your potential audiences: those readers in your own group, those in close proximity to your group, audiences elsewhere in the organization, and audiences outside the organization. In the identification and consideration of each as presented below, you will gain valuable information.

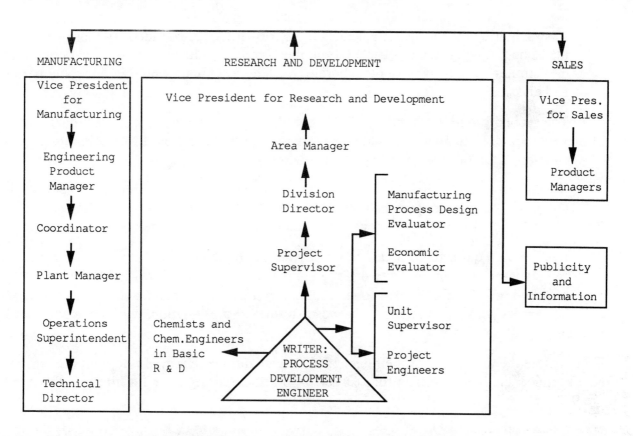

Figure 9.3 Complex Audience Components for a Formal Report by a Chemical Engineer on a Process to Make a High Purity Compound (from John C. Mathes and Dwight W. Stevenson. *Designing Technical Reports: Writing for Audiences in Organizations*. Indianapolis: ITT Bobbs-Merill Educational Publishing Company. 1976. 14)

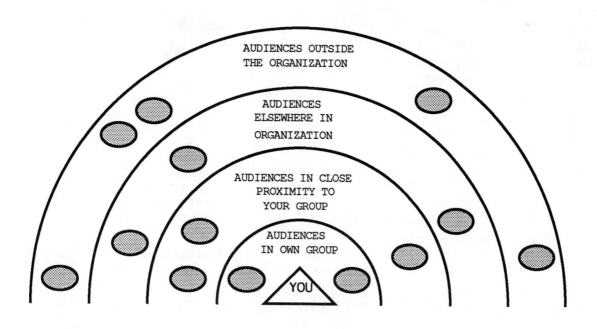

Figure 9.4 Egocentric Organization Chart (from John C. Mathes and Dwight W. Stevenson. *Designing Technical Reports: Writing for Audiences in Organizations*. Indianapolis: ITT Bobbs-Merill Educational Publishing Company. 1976. 15)

- In anticipating the needs of audiences in your own group, you identify those people in the same project group or office. Often, these are individuals with technical competencies similar to your own. In the laboratory, such individuals would include you partners and classmates.

- The second group is comprised of those with whom you normally interact outside of your office. Here may be found managers in your organization. In the laboratory, your course instructor might fall into this audience.

- Readers in the third group are those who are distant from you within the organization: the public relations department, sales, the legal department, purchasing, and so forth. In an educational setting, your department chairman would be a member of this audience.

- Readers in the fourth group are those beyond the organization. They may work for the same company in another city or for an entirely different organization. In an educational setting, non-specialist professionals, such as humanities or mechanical engineering professors who might attend your end-of-term oral presentation, would be considered here.

To illustrate the usefulness of their egocentric organizational chart, Mathes and Stevenson asked a chemical engineer working for a large corporation to design one for himself. The result is Figure 9.5.

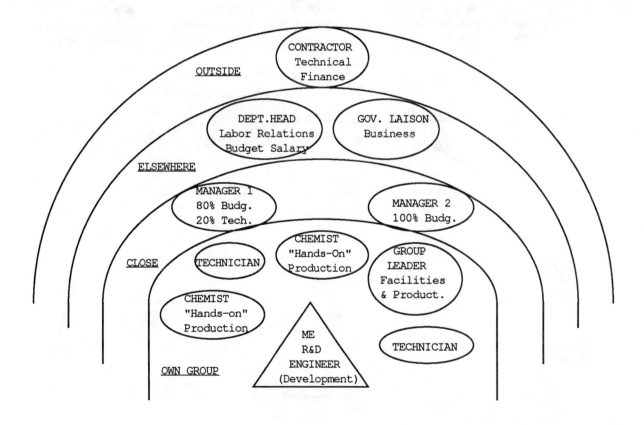

Figure 9.5 Actual Egocentric Organization Chart of an Engineer in a Large Corporation (from John C. Mathes and Dwight W. Stevenson. *Designing Technical Reports: Writing for Audiences in Organizations.* **Indianapolis ITT Bobbs-Merill Educational Publishing Company. 1976. 17)**

You will notice that even those in the chemical engineer's own group have different demands from the writer. A technician, with only two years of college, may have little formal knowledge of chemical engineering. And the group leader is concerned not with technical engineering development but with facilities and production operations. The engineer says that his group leader is already "losing familiarity with the technical material."

The informational needs of those in the second group shift from those of the first group. Often these individuals are concerned with budgets. Mathes and Stevenson's engineer reveals that neither Manager I nor Manager II can follow technical explanations. As the engineer said, both would find it "difficult to return to the lab."

In the last two groups, the audience needs change again. In the third group, the department head is concerned with budget, personnel, and labor relations. In the fourth group, the chemical engineer identified as significant the contractor who, if he is to use subcontractors, will have needs that are both technical and financial.

This first step in audience analysis is useful because it empowers you as your communications "radiate throughout the organization." Practically, an egocentric organizational chart will only have to be designed once to describe your typical reporting structure. It can then be tailored for special reports.

Step 2. Characterize the inividual report readers. As you prepare your egocentric organizational chart, you will begin to think of individual readers, of their needs, knowledge, and biases. Mathes and Stevenson suggest that as you begin to identify individual readers, you should characterize each in terms of operational, objective, and personal characteristics.

- To anticipate the *operational characteristics* of your audience, you want to focus on the daily concerns and attitudes that will enable your reader to react to your report. How, you might ask, will my report affect the reader's role in the organization?

- To anticipate the *objective characteristics* of your audience, you want to focus on specific, relevant background information about the person. Specifically, in reflecting on your audience's technical capabilities you might ask if the reader could participate in a conference on chemical engineering. In your laboratory course, an oral presentation to a *mixed* professional audience requires a high degree of translation. You must communicate with both chemical engineering professors and non-specialists, such as humanities professors.

- *Personal characteristics* are those particular biases—both good and bad—that will enable the reader to be attentive to your information. To anticipate the reader's personal characteristics, think about the personal concerns that the individual has exhibited in the past.

To enable you to characterize your individual report readers easily, Mathes and Stevenson offer the form below in Figure 9.6. Again, this is a practical step in the process of audience analysis: you can build a file of audience characterizations within your organization and add or subtract to this file as the situation changes.

Step 3. Classify audiences in terms of how they will use your report. For each report you write, trace the communication routes on your egocentric organization chart. As you do this, think in terms of the impact your report will have on your organization.

```
┌─────────────────────────────────────────────────────┐
│                                                       │
│   NAME:                       TITLE:                  │
│                                                       │
│   A.  OPERATIONAL CHARACTERISTICS:                    │
│                                                       │
│       1.  His role within the organization and        │
│           consequent value system:                    │
│                                                       │
│       2.  His daily concerns and attitudes:           │
│                                                       │
│       3.  His knowledge of your technical             │
│           responsibilities and assignment:            │
│                                                       │
│       4.  What he will need from your report:         │
│                                                       │
│       5.  What staff and other persons will be        │
│           activated by your report and him:           │
│                                                       │
│       6.  How your report cound affect his role:      │
│                                                       │
│   B.  OBJECTIVE CHARACTERISTICS:                      │
│                                                       │
│       1.  His education - levels, fields, and years:  │
│                                                       │
│       2.  His past professional experiences and roles:│
│                                                       │
│       3.  His knowledge of your technical area:       │
│                                                       │
│   C.  PERSONAL CHARACTERISTICS:                       │
│                                                       │
│       Personal characteristics that could influence   │
│       his reactions—age, attitudes, pet concerns, etc.│
│                                                       │
└─────────────────────────────────────────────────────┘
```

Figure 9.6 Form For Characterizing Individual Report Readers (from John C. Mathes and Dwight W. Stevenson. *Designing Technical Reports: Writing for Audiences in Organizations.* **Indianapolis: ITT Bobbs-Merill Educational Publishing Company. 1976. 20)**

After you have traced the communication routes, you will have a better idea of how your readers will use your report. Mathes and Stevenson suggest a three-tier classification of audiences:

Primary audiences— those who make decisions or act on the basis of the information in your report.

Secondary audiences— those who are affected by the information in your report.

Immediate audiences— those who transmit the information found in your report to others in the organization.

The primary audience is those for whom the report is intended. If, for example, you are a research chemical engineer at Celeanese Corporation who has found a way to increase the resistance of cellulose triacetate fiber (Arnel) to glazing at high ironing temperatures, your audience may consist of multiple readers who are both in close and distant proximity to you in the corporation: both the technical and sales groups may be interested in your discovery. When you generate a Proposal Request for Funding (see Chapter 11) based on your experimental results, you are writing to someone who will decide whether to fund your work to its logical completion. That person will make a decision based primarily on your report.

While the primary audience makes decisions based on your report, the secondary audience is affected by your report. To return to the example from the Celeanese Corporation, you should realize that a secondary audience of your report will be the line manager who, if your new process is adopted, may have to make adjustments in production. Another example is the student intern who will employ the User's Manual (see Chapter 11) you produce based on one of your experiments.

The immediate audience should not be confused with the primary audience. In many cases, the immediate audience might be your supervisor or another mid-level manager. The group leader of research and development at Celeanese Corporation may accept your proposal for increasing the ironing temperatures of Arnel, yet your primary audience may, in fact, be your group vice-president who will make the decision to put your discovery into production. In your laboratory course, your instructor can be considered part of the immediate audience for your report.

To enable you to more readily recognize your readers after you have characterized and classified them, Mathes and Stevenson offer the matrix for audience analysis found in Figure 9.7. At a glance, this matrix will reveal what information you have and what information you will need to correctly determine the identity and needs of your audience.

On the one hand, as Mathes and Stevenson correctly believe, you address a definable audience when you write. You anticipate the audience's needs and seek to fulfill them. Yet, in a very real way, you also invoke your audience. That is, you create in your document a role for the reader to play. The better writer you are, the more you can use your document to project and alter the role of the reader. To return to your discovery at Celeanese Corporation, you would want to write your report so that it is neither so technical that it alienates the group vice-president nor so sales-oriented that you begin to sound like a member of the marketing division. Perhaps you will select a technical translation memo from Chapter 11. In striking a balance by using such a reporting structure, you not only address the reader's needs but you actually control the perception of those needs.

In realizing that you both address and create your audiences, you can recognize the need for a sophisticated process of composing your documents. It is to that process that we will now turn.

The Obligation of the Writer: A Question of Process

Your writing process has most likely been formed in school and college. No doubt, you have experience in answering short-answer test questions and essay questions under tight time constraints. You have experience in composing research papers in which you develop a thesis, find supporting sources, and write up your twelve-page term paper a day or two before it is due. In the laboratory, you have perhaps paid more attention to the technical aspects of your report than to the communicative aspects, hoping against hope that the instructor will praise your technical ability and fill in the blanks left by your failure to communicate.

In many ways, the academic world is forgiving of its undergraduates; if you fail to perform, there is always another chance. No such case exists in the non-academic world you may face upon

	OPERATIONAL	OBJECTIVE	PERSONAL
CHARACTER-ISTICS / TYPES OF AUDIENCES			
PRIMARY	①1	④4	⑦7
SECONDARY	②2	⑤5	⑧8
IMMEDIATE	③3	⑥6	⑨9

Figure 9.7 Matrix for Audience Analysis (from John C. Mathes and Dwight W. Stevenson. *Designing Technical Reports: Writing for Audiences in Organizations.* **Indianapolis:ITT Bobbs-Merill Educational Publishing Company. 1976. 23)**

graduation. Here, your documents will influence the lives of individuals within a complex organizational structure. Here, writing counts in ways that you have not imagined.

To address this new environment, you will need to re-think your writing process. Based on recent research in technical communication and based on our experience in our own undergraduate laboratory, we offer the writing process depicted in Figure 9.8.[21]

Let us review this process in detail:

Inventing. This is the process of discovery. In thinking about the audience for your document, you should explore the audience analysis method discussed above. You should prepare an egocentric organizational chart, characterize individual report readers, and classify your readers in terms of how they will use the report. This method will help you tailor the document to the needs of your audience. In fact, the purpose of your document is interrelated to the type of document you are creating.

When given a writing assignment, experienced writers on the job frequently go to the file cabinet and pull out other similar documents in use in the organization. Thus, a chemical engineer assigned to write a briefing memo frequently will seek another similar document, analyze its purpose, format, and content, and proceed from there. At least in part, the writer can infer the purpose of the document by analyzing the document itself. This point is important for you to remember: if you are assigned to write a document, one of the best

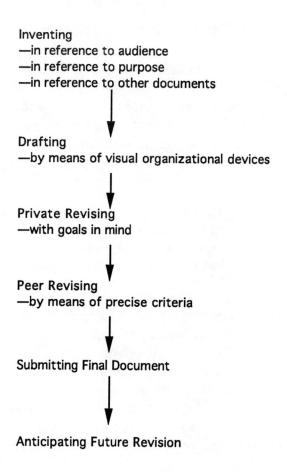

Inventing
—in reference to audience
—in reference to purpose
—in reference to other documents

Drafting
—by means of visual organizational devices

Private Revising
—with goals in mind

Peer Revising
—by means of precise criteria

Submitting Final Document

Anticipating Future Revision

Figure 9.8 A Writing Process for Upper-Level Chemical Engineering Students

things you can do is locate a previous document that was well-received by your organization. We have thus provided the extensive exhibits in Chapters 10 and 11 so that you can study effective examples of the reporting structures you will encounter in chemical engineering.

Drafting. Remember the outline from grammar and high school, complete with Roman numerals, capital letters, lower case letters, and numbers? You might also have had difficulty with those highly structured devices: you could never figure what ideas went where, so you usually wrote your essay and then filled in the outline afterwards so that the teacher would get the required outline.

Why did the method fail? Quite simply, the process of planning was too detailed; you would not know about the details of your paper until you actually wrote it because, as we proposed above, writing shapes your thinking. In fact, this was why the outline was so easy to fill in after you wrote your paper. Research, however, reveals that experienced writers use a system of bullets to sketch out the major ideas and then go back to these major ideas to fill in the details. As you draft your document, simply let the ideas flow. During this stage of the writing process, do not allow your thoughts to become derailed by concern for correctness of grammar, syntax, or spelling. There will be time for editing in the revision phase of the process.

Private Revising. In this phase, you should review the decisions you made in the invention stage of the process. As you recall your audience for the document, your document's purpose, and its similarity to other successful documents, you might want to establish goals for revision. Do you want to identify with the reader so that your ideas will be received favorably? If you are writing for a technical audience, your data should be correct, so the goal of accuracy will be important. After you have made a list of these goals, review the document to see if you have achieved them. If you have not, perform further revision. In addition, at this stage you might want to make your first pass at revision for sentence clarity and grammatical and mechanical accuracy.

Peer Revising. In Chapter 3 you encountered the subjects of collaborative work and assessment. In Chapters 10 and 11, you will find a number of criteria sheets that address the reporting structures. These assessment sheets can be used with your classmates and instructor for collaborative work so that they can review your documents before they are submitted.

Why use the kind of review discussed in Chapter 3? Others can see problems that you may have missed and, more importantly, they can confirm if the document will be effective. Why use criteria sheets? It is our experience that, unless you ask someone to read with a purpose, their comments will be so general ("Yea, this looks good") that they will be useless. However, if you ask a member of your group or an instructor to respond to something about your document that concerns you—"Can you understand what this memo is about in the first two sentences?", "Is there any place that you get lost or confused?", "Will my tone upset the reader so much that my message will get lost?"—you will get a much more elaborate and useful response.

Submitting the Final Document. If you view the creation of your document as a kind of experiment, you have now taken two runs at it and you should now have some great data. Look this data over and incorporate it into your report. To extend our analogy, just as there are levels of data we have described in Chapter 4, so there are levels of comments. Some will be proofreader's comments, others questions of accuracy of data. There may be comments on tone and comments on format. Your job, just as you must group levels of empirical data, is to group the comments you receive. After these are grouped, read and revise your document with one group in mind at a time. For example, do not read for sentence length at the same time you are checking the accuracy of a table. As you prepare to submit your final document, then, read and correct it in phases.

Anticipating Future Revisions. In school, an essay is often written, submitted, graded, and discarded. This is not the case in the organization. Research tells us that the life cycle of documents ranges from those that are immediately tossed in the wastebasket to those that remain in use indefinitely. A briefing memo on your work to overcome the poor resistance to moisture of cellulose acetate may be retained only until the next memo is submitted, while a procedural manual for a Podbielniak countercurrent rotating contactor may be in existence as long as the centrifugal extraction equipment is used. A request for funding to purchase the Podbielnaik may exist only as long as a budget decision is pending, while an oral presentation proposal on the new product that was crated as a result of your research with cellulose acetate may be made for various audiences again and again. In many cases, revision is a never ending process.

This writing process is the one that we have used with our own students to enable them to create the exhibits you are about to study. We believe the model in Figure 9.8 is useful for three reasons.

First, it is based on research in the field of communications, so it is not merely another clever academic idea that bears no relationship to reality. Again and again, current research in communications is showing that the writing process of successful organizational writers is based on the kinds of realistic contingencies we have built into our model.

Second, we believe our model is useful in that it demystifies the writing process. On one hand, we have found that our students are often disabled by a mistaken notion of writing that depicts the isolated scholar sitting in his study, sipping brandy, waiting for inspiration to strike. On the other hand, our students ironically often seem to have a model of the writing process that is similar to that of novice writers struggling to master the level of basic literacy.[22] These novice writers spend a great deal of time on mechanics and grammatical correctness, do little planning and self-analysis of their writing, and do not use goals to revise their work. These two inappropriate models— the solitary creative writer and the struggling basic writer—disable you. The model we present, instead, serves to empower you because of its reality and its clarity.

Third, we believe our model is useful in that it exemplifies the kind of critical thinking discussed in Chapters 1 and 3. Indeed, the composition process of inventing, drafting, revising, and submitting a document closely resembles the engineering process of defining a problem, formulating an experiment, observing the evidence, and presenting results (see Figure 3.4). Because our model composing process so resembles the engineering process, our students have found great motivation to adopt a new method of writing about chemical engineering. We hope that you, too, will discover this motivation.

Notes and References

1. For an excellent review of these topics, see Jackson J. Spielvogel, *Western Civilization*, New York: West Publishing Company, 1991. Chapter 17, "Toward a New Heaven and a New Earth: The Scientific Revolution and the Emergence of Modern Science," pp. 562–592. See also Stephen F. Mason, *A History of the Sciences,"* New York: Collier, 1962. For a discussion of Stephen Smale's horse shoe, see James Gleick, *Chaos: Making a New Science*, New York: Viking, 1987, pp. 50–53.

2. B. Latour, " Visualization and Cognition: Thinking with Eyes and Hands," *Knowledge and Society: Studies in the Sociology of Culture Past and Present*, 6 (1986), pp. 1–40.

3. Latour, p. 3.

4. Bruno Latour and Steven Woolgar, *Laboratory Life: The Construction of Scientific Facts*, Princeton: Princeton University Press, 1986.

5. Latour and Woolgar, p. 243.

6. Walter Ong, *Orality and Literacy: The Technology of the Word*, London: Methuen, 1982. See especially pp. 78–116.

7. For more on the ways that the context of a discourse community determines the way that we communicate, see the following: Linda Flower, "Cognition, Context, and Theory Building," *College Composition and Communication*, (October, 1989), pp. 282–311.

8. For more on literacy as a means of empowerment, see the following: Paulo Freire, *Pedagogy of the Oppressed*, Trans. Myra Bergman Ramos, New York: Seabury Press, 1968; W. Ross Winterowd, *The Culture and Politics of Literacy*, New York: Oxford University Press, 1989; Stephen. R. Graubard, ed., *Literacy in America, Daedalus: Journal of the American Academy of Arts and Sciences*, 119 (Spring 1990); Andrea A. Lunsford, Helen Moglen and James Selvin, eds., *The Right to Literacy*, New York: Modern Language Association of America, 1990.

9. The discussion below is adapted from Norbert Elliot, Maximino Plata, and Paul Zelhart, *A Program Development Handbook for the Holistic Assessment of Writing*, Baltimore: University Press of America, 1990.

10. Primo Levi, *The Periodic Table*, Trans. Raymond Rosenthal, New York: Schocken Press, 1984. We strongly recommend that every chemical engineering student read this book.

11. Levi, pp. 224–225.

12. For more on the early development of technical writing, see Michael G. Moran, "The History of Technical and Scientific Writing," in M. G. Moran and D. Journet, eds., *Research in Technical Communication: A Bibliographic Sourcebook*, Westbrook, CT: Greenwood Press, 1985, pp. 25–38.

13. Robert. J. Connors, "The Rise of Technical Writing Instruction in America," *Journal of Technical Writing and Communication* 12 (1982), pp. 329–52.

14. Edwin T. Layton, Jr., *The Revolt of the Engineers: Social Responsibility and the Making of the Engineering Profession*, Baltimore: Johns Hopkins Press, 1986.

15. Connors, pp. 329–352.

16. For a technical writing textbook incorporating this definition, see Herman Estrin and Norbert Elliot, *Technical Writing in the Corporate World*, Los Altos: Crisp, 1990. See also the following: N. Elliot and P. Zelhart, "Hermeneutics and the Teaching of Technical Writing," *The Technical Writing Teacher*, 17 (1990), pp 150–164; Norbert Elliot and Margaret Kilduff, "Technological Writing in a Technological University: Attitudes of Department Chairs," *Journal of Technical Writing and Communications*, 24 (1991), pp. 411–424; Norbert Elliot, "Teaching Technical Writing: Notes Toward a New Paradigm," *Technology Studies*, 8 (1989), pp. 5–10. For further discussions of the significance of technical writing, see the following: Carolyn R. Miller, "A Humanistic Rationale for Technical Writing," *College English*, 40 (1979), pp. 610–17; David Dobrin, "Is Technical Writing Particularly Objective?" *College English*, 47 (1985), pp. 237–51.

17. Robert M. White, "Technological Competitiveness and Chemical Engineering," *Chemical Engineering Progress* (January 1988), pp. 24–26.

18. J. J. Peters and R. H. Waterman, Jr., *In Search of Excellence*, New York: Harper and Row, 1982; Siegfried Streufert and Robert W. Swezey, *Complexity, Managers, and Organizations*, Orlando: Academic Press, 1986. Especially pp. 73–89.

19. Isaac Asimov, "The Future of Chemical Engineering," *Chemical Engineering Progress*, (January 1988), pp. 43–49.

20. The following discussion of audience analysis, including the figures, are taken from John C. Mathes and Dwight W. Stevenson, *Designing Technical Reports: Writing for Audiences in Organizations*, Indianapolis: ITT Bobbs-Merrill Educational Publishing Company, 1976. While a second edition was published by Macmillian in 1991, only the 1976 edition includes the case study of the chemical engineer.

21. Jack Selzer, "The Composing Process of an Engineer, "*College Composition and Communication,* 34 (1983), pp. 178–187; Glenn J. Broadhead and Richard C. Freed, *The Variables of Composition: Process and Product in a Business Setting,* Carbondale: Southern Illinois University Press, 1986; D. A. Winston, "Engineering Writing/Writing Engineering," *College Composition and Communication,* 41 (1991), pp. 58–70.

22. Robert B. Kozuma, "The Impact of Computer Based Tools and Embedded Prompts on Writing Process and Products of Novice and Advanced College Writers," *Cognition and Instruction,* 8 (1991), pp. 1–27.

Six Assignments for Sophomores and Juniors: Investigating Professional Communication

Assignment 1

Return to Figure 9.1. What have been your experiences with each of the types of literacy? Have you had opportunities to try your hand at each of these levels? In four paragraphs, reflect on your experiences with basic, academic, disciplinary, and epistemological literacy.

Assignment 2

Look through a copy of the undergraduate catalogue at your university and discuss the following with your classmates: Beyond freshman composition, what kinds of writing courses are offered? What fields require that their students take writing courses other than freshman composition? How do these upper-division writing courses seem to be designed? Should you take one of these courses even if it is not required by your major?

Assignment 3

The communications model in Figure 9.2 is based, in part, on the discussion of technology by Arnold Pacey in *The Culture of Technology* (Cambridge: MIT Press, 1983.) Obtain a copy of the book from your library, and consider the significance of Pacey's interpretation of technology.

Assignment 4

If you are to prepare an oral report in one of your classes, use the Mathes and Stevenson audience analysis method provided in this chapter. Show your figures to your instructor before you make your report to make sure you have anticipated the needs of your audience.

Assignment 5

Turn to Figure 9.8 and draw a figure of your writing process. How does your figure differ from the one we have provided? What is the source of that difference? After you have drawn your figure, compare and contrast your figure with those of other classmates.

Assignment 6

The detailed exercise below will allow you to study the communications situation of a chemical engineer. By formally studying the kinds of writing and

(more)

Six Assignments for Sophomores and Juniors: Investigating Professional Communication

oral presentations a chemical engineer commonly does, you will have a better idea of the kind of abilities you will need after graduation.

For additional help with this assignment, refer to Chapter 10 and the discussion of the field research paper.

Case Study of A Chemical Engineer: Communications in the Work Place

Background to the Exercise: What kinds of writing and presentations do chemical engineers perform in the work place? Do they feel their education prepared them for their present tasks they must daily face when they are asked to present their ideas? What is the relationship of writing ability to career advancement? What are common constraints to writing in the field? How can these be overcome? What kinds of documents do chemical engineering students find themselves writing five years after graduation? Ten years? The field interview will allow you to explore some of these questions.

To begin, spend some time selecting a model chemical engineer in your profession. Obviously, the study you perform will only be as good as your subject. Your instructor will be able to help you identify an appropriate candidate for your interview. Then, define the kinds of questions you want to ask regarding the communications the individual must perform.

The organization of the report you will write will help you plan the interview.

Organization of the Report: Your report should follow this basic pattern:

> *Lead*: Here you set the stage for what will follow. This opening section should provide the reader with a sense of your interview's precise research interest and the design of your paper.

> *Background*: In this section of your case study, you should summarize the education and work experience of the chemical engineer you have interviewed. As well, you should point out those background characteristics which are most significant in relation to your research interest.

> *Methodology*: Here you will tell your readers about the procedure you followed in conducting the interview. In addition, illustrate how each part of your research interest ties into the questions you asked the

(more)

Six Assignments for Sophomores and Juniors: Investigating Professional Communication

individual you interviewed. Answers provided in this section of your report will allow your reader to understand how you obtained the information you are presenting.

Analysis: This is the most important section of your paper. Here you present the responses you obtained from your interview and provide your comments on those responses. What conclusions can you draw? What are the limits of those conclusions?

Portrait of a Chemical Engineer: At the conclusion of this section, readers should feel that they have been presented with a brief, analytical portrait of a chemical engineer's responses to your research interest.

Advice to College Chemical Engineering Instructors: Based on what you have found in the interview, do you have any suggestions for your faculty about enhancing the communications curriculum at your university for majors in your field?

Directions for Further Research: Since this is perhaps your first case study of this type, explain what you would do differently if you performed another study? What methodology would provide more significant information? What kinds of studies could be conducted to enhance and extend the information you have presented?

Criteria for an Effective Report: An effective case study will have many of the following features:

- A well-focused, precise research interest for the interview

- A procedurally-oriented statement of the methodology

- A lucid summary of answers

- Subtle, tentative analysis of those answers

- A precise portrait of the communication situation of the chemical engineer

- Tentative recommendations for curriculum revision in communications at your university

- Well-formulated directions for further study

Tips for An Effective Oral Presentation

Piero Armenante
Professor of Chemical Engineering, Chemistry,
and Environmental Science

Except for the first sentence and the concluding sentence, don't memorize the content of your presentation. (You are not Robert Frost giving a poetry reading.)

Don't read your talk. (You are neither Lincoln nor Douglas.)

Of you are using slides, arrange them in the carousel yourself prior to the presentation. Check that they are in the right order and in the correct position. If you are using overhead transparencies, follow the same procedure. (If you have 15 minutes to present a paper on thermodynamics, and you must stop and shuffle your visuals, you are dead.)

Before you are called to the podium, sit (calmly) in your chair. Do not move until the chairperson mentions your name. However, be prepared. Have the sets of transparencies ready in a folder in your hand or have the carousel with your slides already placed in the projector.

If you use a slide projector, make sure you know how to use the remote control and focus mechanism. If you use an overhead projector try to learn beforehand how it works. (Engineers really look like idiots if they can't work the equipment.)

Make sure you have a pointer. If a laser pointer is available make sure you know which button to push to activate it.

When you are finally called, take a deep breath, move to the podium, and arrange your material before you start talking. Place the first overhead on the projector (without necessarily turning it on), refamiliarize yourself with the projector, its focus mechanism and switches, examine the laser pointer, adjust the microphone to your level, and so on.

Make a mental note of the time at which you are starting your talk so that you know by when you have to complete it. In most

(more)

cases you will have only a limited amount of time to deliver your talk. Stay within this limit! (Practice beforehand to make sure that you don't run into the other presenters' time.)

Once you are ready, and only then, you should begin to speak. Begin by greeting the audience and the chairpersons. If, for whatever reason, your name was not announced, introduce yourself and mention your organizational affiliation.

Put on the first slide and begin your talk with the first sentence in your presentation. (This is the one that you memorized just in case your nerves get the better of you.)

Use an annotated set of hard copies of your slides or overheads to remind yourself of your talk. (Again, remember that you are speaking from visual representations, not from a printed paper.)

Even to experienced audiences, give a detailed introduction to the subject. In the vast majority of cases people will not be familiar with your subject, and you will have to lead them into the core of your talk. In most cases you will have in front of you an educated and intelligent audience. If you explain something new and even complicated to them in a clear and logical fashion, they will understand. (If you build it, they will come.)

Never leave a slide alone. If you put a slide on, go through it before turning to the next one. Do not confuse the audience by talking without elaborated reference to the slide. The people will not know whether to listen to you or figure out the content of the slide on their own. (You don't want members of your audience to interpret your data for you, especially if they see trends that you are not prepared to discuss.)

Allow a sufficient amount of time to go over each slide. A typical rule of thumb is 1 to 2 minutes per slide.

Avoid putting too many concepts/equations/figures/ tables/ numbers on the same slide. Do not overwhelm the audience. (See the sample overheads in Chapter 11.)

(more)

Conclude your talk by very briefly recapping the most important point of your presentation. Then say that this concludes your talk and thank the people in the audience for their attention.

In formal talks the end of the talk is (typically) followed by an applause.

After the applause, ask if there are any questions. Do so as if you are inviting challenging questions. (Do not look like someone who wants to run away or hide behind the podium. It won't help. They will hunt you down regardless.)

If you do not understand a question, do not be afraid to say so. Most of the time you will be able to answer the question appropriately. (After all, chances are you will know your project better than anybody else in the audience.) However, occasionally you may genuinely not have a good answer. If you don't have even the foggiest idea of what a reasonable answer could be, then say that the point raised was a good one, and mention that you will think about it and go back (hopefully with an answer) to the person asking the question after the session is over. Give yourself a good chance to answer appropriately; however, do not be afraid to say that you have never thought about the problem if this is the only way out.

(Never, never try to fake an answer.)

Think before answering the question. People are always impressed by speakers who ponder on their questions, since this thoughtful pause implies that the question was challenging and the questioner was smart. So take advantage of this nutty phenomenon and think a few seconds before saying something.

10 Forms of Communication for Students of Chemical Engineering: In the Classroom

Reading and writing have made Western technology, science, philosophy, literature, and theology possible.

W. Ross Winterowd, *The Culture and Politics of Literacy*

Introduction

Just as all ideas exist in a context, so too do the ways that we document our ideas. It is to the documentation of ideas that we now turn.

This chapter contains strategies for presenting ideas. Students of chemical engineering, we have found, must think and write in the realistic context of a classroom while, at the same time, working in the laboratory. In Chapter 11, we will turn to forms of writing associated with the laboratory. In this chapter, we will present forms of writing common in classes you'll take in the humanities and the social sciences.

What You'll Find in the Discussion of Each Type of Academic Reporting Structure

In the following pages we will present guides that will enable you to produce four reporting structures commonly found in the classroom:

- lecture notes,
- the experiential paper,
- the field research paper, and
- the historical research paper

In the presentation of each of these kinds of documents you will find:

— a brief *introduction* to the reporting structure,

— a section on *organization* to guide you in preparing the report

— a brief *description* of the assignment on which the reporting structure is based,

— a sample of the competed report, called the *exhibit*,

— a series of *assessment criteria* by which the report may be judged, and

— *questions for discussion* about the exhibit which generally refer to the annotated instructors' comments.

A Note on the Creation of the Exhibits

The exhibits were created by our students in a two-semester sophomore and junior Science, Technology, and Society class joined to introductory classes in chemical engineering. All papers were written under the writing process model provided in Chapter 9. All papers were drafted by our students, revised under our supervision, and then resubmitted for a final grade. We have selected what we believe to be solid sample student papers which successfully fulfilled the various assignments and aptly illustrate a rich array of critical thinking abilities. For the sake of brevity, in some of the exhibits we show only those portions which render the paper unique and relevant to our purposes here.

Taking Lecture Notes

Introduction

If information is not processed in some way, it is gone within 30 seconds. You can rehearse it, you can write it down, but it won't stay in your mind by itself.

So if your history instructor is talking about an important manufacturing issue in the Civil War or your calculus instructor is referring to an important formula not found in the textbook, you have no more than 30 seconds to process the information.

The process of taking notes is essential to your success both in the classroom and in the laboratory.

Organization

Walter Pauk's Cornell Note-taking System has been, for over forty years, the best system for recording class notes in a hand-written textbook.[1] The exhibit provides our version of Pauk's well-known system. The features are simple:

Use an 8.5" by 11" page. Down the left side, draw a vertical line; the space to the left of this line will be Block 1. How much space you need is dependent upon the way you take notes in your field. Pauk recommends that the line be drawn 2.5" from the left of the page, but that seems rather close to us. In Exhibit 1, we actually give a good deal of space (4" or so) to Block 1. This space is for jotting down ideas as they are presented.

The space to the right of the vertical line is Block 2. Pauk recommends that you take your notes in Block 2. We have found that this space can also be used for cross-references to other notes and materials when you later review your notes.

Draw a horizontal line at the bottom of the page about 2" from the edge of the paper. Here you can draw figures or copy formulas. We call this Block 3.

While this system is elementary, our students have found it enormously helpful. If you arrive for a lecture or a meeting with such pages pre-designed, there is little chance that you will miss much information.

Description of the Lecture

The completed notebook page that we have exhibited captures a class discussion of the origin of the American Institute of Chemical Engineers. This type of lecture is important in helping you understand the rich history of your field.

Block 1: To Capture Ideas **Block 2: To Elaborate on Ideas**

In this space, you can jot down ideas as they are presented. We suggest that you bullet your ideas.

In this space, you could build on the ideas you jotted down in Block 1. You can, for example, make notes here on a report mentioned during a meeting.

In this space, you can draw a figure from an overhead or copy a formula.

Block 3: To Sketch Figures and Copy Formulas

198

Date: 4/12/xx
Lecture: The History of the AIChE:
1908–1930; Guest Lecturer in ChE 226:
Dr. John Lane

Sources of Information for Further
Reading: Terry Reynolds, *Seventy-Five
Years of Progress* (1983)
T. Reynolds "Defining Professional
Boundaries," *Technology and Culture*,
Vol. 27, 1986, pp. 694–716.

- early history dominated by problems:
- definition of the field
- legitimacy of ChE as independent field
- rival fields:
- industrial chemistry
- applied chemistry
- mechanical engineering
- Society formed in 1908; 40 members;
  today over 50,000 members
- Early interest by AIChE in education
- Early programs in chemical engineering at MIT, The University f Pennsylvania, Tulane, and Wisconsin
- Big moment of self-definnition for
  AIChE: 1915, A.D. Little's report to the
  president of MIT in which definition of
  unit operations is given. This term
  gave the new field what it needed: a
  way to define its own uniqueness

Get this book from the libray; find out if
we take this journal; if not, inter-library
loan.

All of this sounds like the discussions of
America in the late 19th–century that
we're doing in my hist. of technology
course.

Dates of founding of other organizations
from other fields:
    Amer. Society of Civil Engineers: 1867
    Amer. Inst. of Mining Engineering: 1871
    Amer. Inst. of Electrical Engineers: 1884
    Institute of Radio Engineers: 1912
(From *Technology in America*, pp. 165–
173)

The program at MIT is described by
Ralph Landau and Nathan Rosenberg in
Invention and Technology, Fall 1990, pp.
58ff.
Wisconsin's program by Olaf A. Hougen
in *CEP*, Janu. 1977, 89–ff.

Unit Ops = Part of the Systematization of
America???

Important Founders:

Arthur N. Noyes, William H. Walker, Arthur D. Little, Warren K. Lewis

Important Early Textbooks:

Richards, *Metallurgical Calculation* (1906)

Lewis and Radasch, *Industrial Stochiometry* (1926)

Houghen and Watson, *Industrial Chemical Calculations* (1931)

Assessment Criteria for the Lecture Notes

Below are listed the criteria for well kept classroom *lecture notes*. After reading the exhibit, please circle the response indicating your judgment as to whether the student has exhibited

4 - Superior ability

3 - Competent ability

2 - Somewhat competent ability

1 - Lack of competent ability

1. The notebook page allows for rapid note-taking.

    4   3   2   1

2. The notebook page allows for later reflection and refinement of the lecture.

    4   3   2   1

3. The notebook page allows for the visual display of information.

    4   3   2   1

4. The student has returned to the lecture notes and elaborated on them.

    4   3   2   1

5. The student has used the lecture notes as an occasion for cross-referencing the notes to lectures given in other classes.

    4   3   2   1

Discussion Questions on Taking Lecture Notes

1. How does the sample demonstrate that the student has captured the most important parts of the lecture?

2. What kind of research would the student have had to do in order to complete Block 2 of the notebook?

3. What kind of illustration could have been drawn in Block 3?

4. How is this method of taking notes similar to that method described in Chapter 11.

5. Identify instances of critical thinking in the sample page of lecture notes.

---

[1]Walter M. Pauk, *How to Study in College*, Boston: Houghton Mifflin, 1989. See especially pp. 121-161.

The Experiential Paper

Introduction

The most common type of communication in the undergraduate curriculum is the essay. The papers that are due in most non-technical classes are submitted in this traditional academic reporting structure. (The papers due in your field—chemical engineering—take different forms; these will be addressed in the next chapter.)

Historically, the essay was invented by Michel de Montaigne (1533-1592). His *Essays*, published between 1580 and 1588, were translated into English in 1603. Intensely personal, often whimsical, and frequently digressive, Montaigne's essays exhibited the Renaissance view that the individual was worthy and thus deserved extended treatment. Montiagne's essays are therefore personal. "If the world finds fault with me for speaking too much of myself," he wrote," I find fault with the world for not even thinking of itself."

Today, the essay takes various forms. Often, a popular form you might encounter is the experiential, or personal, essay. Composing this type of paper allows you to explore various subjects by deliberately writing about them in an autobiographical fashion. In a tradition that includes *The Periodic Table*, the great autobiographical study by Primo Levi, the experiential paper affords you the opportunity to develop your own unique voice while examining topics often found in the undergraduate curriculum.

Organization of the Experiential Essay

There are many ways that essays can be organized. The organization below is only one strategy that you might use if you were asked to write an experiential essay explaining your decision to become a chemical engineer.

Introduction

Establish how you came to major in chemical engineering.

In this narrative section, introduce a sense of your academic and personal experiences so that the reader has a sense of you as a student of chemical engineering.

Areas of Present Interest

Identify and explore those areas of chemical engineering that most hold your interest. If, for example, you are interested in the application of polymers to the clothing industry, explain this interest here.

Career Interests

Identify and discuss your career interests.

Discuss issues of the opportunities and responsibilities implicit in your career as a professional engineer.

Ending

Close with a statement of your future professional goals in chemical engineering.

Use this section of the paper to make sure you have identified areas of your profession in which you are most interested.

References

Provide the reader with a list of the books and articles you have used in support of your ideas. (Note: The writer of the exhibit has used the documentation format of the American Chemical Society. A brief summary of that format has been provided for you, immediately following the exhibit.)

Description of the Paper

This paper was written by a sophomore student just beginning a major in chemical engineering.

My Role in Chemical Engineering: A Preliminary Report

Today's world is concerned with the impact of accelerated industrialization and its effect on the global environment. Due to the rise of a global economy, the impact of industrialization on the planet has created major concerns regarding the hazardous effects on the world's environment. Such concerns as acid rain, deforestation, and the mass uses of fossil fuels have caused environmentalists and today's engineers to address these concerns on a global basis. As a team of editors recently wrote in *Chemical Engineering Progress*, chemical engineers clearly are at the forefront of many of today's major public concerns (Cynthia F. Mascone et al., 1991).

An example of the new technology being developed by today's chemical engineers is found in new fuel products. The new gasolines now being researched will cause a cleaner burning of the fossil fuels now used in the combustible automobile engine. The research now being developed by the chemical engineers in discovering new and different types of fuels, I believe, is the beginning of an ever-expanding field which requires many hours of thoughtful research and development. It is this field that I am preparing to enter.

In High School

At the time I was in high school I did not know which profession to follow, but I felt that it should be somewhere in the field of science. In high school I received very high grades in both mathematics and science. My high school counsellor suggested to me that since I did so well in these classes that I should look to continue my education in college in the engineering field.

I had great environmental concerns then, and I believed that I could help the world and its environment through the pursuit of a career in chemical engineering. Eventually, I felt that I could best help clean up part of the air pollution problems through the discovery of new fossil fuels to be used in the combustible engine. I believed that engineer fuel research was one very effective way of reaching my goal of aiding the clean up of air pollution. For that reason, I decided to major in chemical engineering.

The Field

The field of chemical engineering is one that is very hard to define. Today there is no universal definition of chemical engineering accepted by the profession (Felder and Rosseau, 1986). I believe that the profession is so broad-based because of the numerous fields of chemical engineering and that it is difficult to derive a definition suitable to all its different aspects. (Of course, the diversity of the field, as Terry Reynolds explains, has been a part of chemical engineering; in 1908, the date of the founding of the American Society of Chemical Engineering, the field was as hard to pin down as it is today.)

In the beginning I had strictly believed that engineering students learned only chemistry, physics, and mathmatics. Engineering appeared to be a profession which operated strictly by the book. It seemed that to be a successful engineer all a student had to do was to learn the mechanics behind the academic course.

By taking courses, however, I began to see that I would need other courses to carry out the requirements of my profession. I began to see that there was one important ingredient needed in order to determine the correct decision to make. That ingredient was one of ethical and moral standards. It was important for me to develop a personal system of ethics and morality which I would later use in my professional career. The logic behind ethical decision making is thus important to me. Thus, there is more to engineering than the mechanics behind the academic course. The correct ethical and moral decisions also have to be considered in deciding whether the ends justify the means.

Co-Op and AIChE

My present plan includes participation in the co-op program. The program mixes real-life experience while allowing me to obtain the appropriate education in such a manner that I will get to see and apply knowledge in an industrial setting. The advantage of cooperative education is explained by Hughson and Lipowicz:

> U.S. educated chemical engineers are fairly well prepared for their first jobs after graduating from college, although many lack communication skills, knowledge of working in a business environment and the ability to work with others (58).

With the work experience obtained through co-op I will not be put in a position of great difficulty of obtaining a permanent job after graduation since I will have the required work experience so often demanded by industries today. And I will have no regrets about my decision because I think that it is the best choice I am going to make. "Almost all engineers who had co-op (work-study) experience value it highly, and many without such experience wish they had it" (Hughson and Lipowicz, 48).

In addition to co-op I think that the AIChE program is important also. This program also will help me find a job when I graduate. However, it offers more than that. It will help my financial difficulties by setting up scholarship programs, contexts, paper writing awards, and many other programs. Also, it will give me the opportunity to meet people who are already in my field who can give me advice and guidance through my years of college.

Future Goals

My goal of the future is to obtain a job in the petroleum industry in order to attempt to change the present and future fuels being used in the combustible engine. I agree that engineers must discover cheaper and better production methods (Landau).

In order to begin to start upon this long-range goal, my first objective is to receive my bachelors degree from my university and then get a permanent full-time job in the petroleum industry. When I obtain this I would further my education my taking night courses in order to get my masters degree in chemical engineering. Beyond obtaining my masters degree, I would also like one day to get my Ph.D. degree in chemical engineering. I believe that the greater level of education is the key to my success in obtaining the knowledge necessary to discover eventually the fossil fuels of the future. These fuels will have an environmental impact and make the air we breathe a little more cleaner and make us a little healthier.

References

Felder, R.; Rosseau, R. *Elementary Principles of Chemical Processes*, 2nd ed.; John Wiley: New York, 1986.

Hughson, R.; Lipowicz, M. *Chem. Eng.* **1983**, 19 Sept., 48-60.

Landau, R. *Chem. Eng. Prog.* **1989**, Sept., 25-39.

Mascone, C. F.; Santaquilani, A. G.; Butcher, C. *Chem Eng. Prog.* **1991**, Oct., 73-82.

Reynolds, T. R. *Seventy-Five Years of Progress: A History of the American Institute of Chemical Engineers, 1908-1983*; AIChE: New York, 1983.

Documentation System: ACS

Documentation System
The American Chemical Society
(Adapted for Undergraduate Student Use)

Journal Articles

Armenante, P. M. ; Kirwan, D. *Chemical Engineering Science* **1989**, 2781-2796.

Books

Bockris, J. O.; Reddy, A. K. N. *Modern Electrochemistry*; Plenum: New York, 1970; Vol. 2, p 132.

Chapters in Edited Volumes

Trattner, R.; Sedlack, S. In *Encyclopedia of Environmental Control Technology*; Cheremisinoff, R. N., Ed.; Gulf Publishing: Houston, 1990; Vol. 4, pp 201-227.

Material Presented Orally

Barat, R. "Characterization of the Chemistry/Mixing Interaction in a Toroidal Jet Stirred Combustion"; Presented at the AIChE Annual Meeting, Chicago, IL, 1990.

For more information about the ACS system of documentation, see the following:

Dodd, J. S., Ed. *The ACS Style Guide*. American Chemical Society: Washington, D.C., 1986.

Assessment Criteria for the Experiential Essay

Below are listed criteria for the *experiential essay*. After reading the exhibit, please circle the response indicating your judgment as to whether the student has exhibited

4 - Superior ability

3 - Competent ability

2 - Somewhat competent ability

1 - Lack of competent ability

1. There is an examination of the relationship between the writer's experiences and the writer's decision to pursue chemical engineering as a major.

    4   3   2   1

2. There is an identification and extended discussion of the areas of chemical engineering that most interest the writer.

    4   3   2   1

3. There is an identification and extended discussion of the demands and challenges of a career in chemical engineering.

    4   3   2   1

4. The writer has used a variety of secondary sources in order to support the points made in the paper.

    4   3   2   1

5. The writer has used a recognized form of documentation so that readers may easily find the sources for their own reference.

    4   3   2   1

Discussion Questions for the Experiential Essay

1. Has the writer provided a lead that establishes the tone and direction—the essential nature—of the essay?

2. Has the writer used secondary sources that are well chosen? How do these sources contribute to the paper?

3. Does the ending suggest that the writer has stepped back and given due final consideration to the ideas presented in the essay?

4. Are there signs of careful craft in the essay? Note these and discuss them.

The Field Research Paper

Introduction

The field research paper is often found in the social science curriculum. That is, you will encounter such papers in such courses as sociology, political science, psychology, economics, history, communications, and geography, as well as in courses in science, technology, and society. In these courses, human social behavior is subjected to the same kind of scientific study as the engineer undertakes in the laboratory.

In a field interview—one kind of field research assignment—you are able to construct a carefully designed set of questions, ask them to a carefully selected respondent, and analyze the answers. In a field study of a chemical engineer, you might ask these types of questions: What kinds of jobs do chemical engineers perform in the workplace? Do they feel their education prepared them for their present tasks they must daily face? What is the relationship of postgraduate education to career advancement? What are common constraints to success in the field? How can these be overcome? What kinds of work do chemical engineering students find themselves doing five years after graduation? Ten years? The field interview will allow you to explore some of these questions.[1]

Interviewing a practicing professional in your field—a typical assignment—will provide you an opportunity to know more about the kinds of work you'll be performing after graduation and how you might best prepare for that work.

A random, informal interview will, however, do you little good. Instead, you should carefully plan your field study based on a single focus. This focus about your field—a study of the role of research and design in computer engineering, a study of the role of pollution prevention in the formulation of a company's policy decisions—will help you narrow your interview. In the accompanying exhibit, we begin by providing a cognitive complexity map that will allow you to narrow your case study. This complexity map is the basis for the sample student paper that is also provided.

Organization of the the Field Research Paper

A common organization of the field research paper based on a personal interview with a practicing professional is provided below.

Lead

Set the stage for what will follow. This opening section should provide the reader with a sense of your precise research interest of the interview and the design of your paper. In the exhibit the student begins his case study with precise statements of the focus (often termed the concept of interest). With this kind of focus, the student demonstrates a characteristic of a well-planned case study. Note that the student also presents a lead that sets the organizational pattern of his case study.

Background

Summarize the education and work experience of the person you have interviewed. As well, point out those background characteristics which are most significant in relation to your concept of interest. (Hint: Don't spend time on this information during the interview; if you plan ahead in setting up the interview, simply ask for a resume. With this information in hand, you can focus your background questions more precisely.)

Methodology

Tell your readers about the procedure you followed in conducting the interview. In addition, illustrate how each part of your research interest ties into the questions you asked the individual you interviewed. Answers provided in this section of your report will allow your reader to understand how you obtained the information you are presenting.

Analysis

This is the most important section of your paper. Present the responses you obtained from your interview and provide your comments on those responses. What conclusions can you draw? What are the limits of those conclusions?

Portrait

At the conclusion of this section, readers should feel that they have been presented with a brief, analytical portrait of the individual's responses to your concept of interest.

Suggesstions about Curriculum

Based on what you have found in the interview, offer any advice about enhancing the curriculum at your university for majors in your field. Note that this is advice, not criticism.

Directions for Further Research

Provide speculation regarding what you would do differently if you performed another study. What methodology would provide more significant information? What kinds of studies could be conducted to enhance and extend the information you have presented?

References

Provide here the sources you have used to test and refine the information that helped you design your research and analyze its results. A one-page guideline of the documentation format of the American Psychological Association is provided at the end of the exhibit.

Description of the Exhibit

In the exhibit, the student has designed, performed, and written a case study of a practicing chemical engineer. The concept of interest is the way that the university curriculum has prepared the respondent for success in the profession.

The Cognitive Map

Sample Cognitive Complexity Map for Field Research Paper

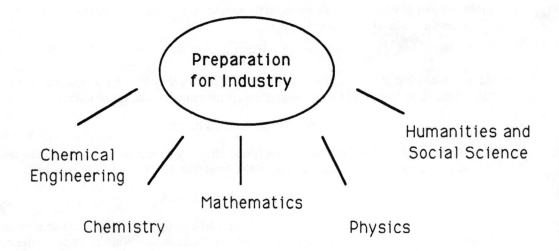

The Preparation of a Chemical Engineer

This case study was designed to explore the academic preparation of a chemical engineer. In this case study I will consider the preparation of a selected chemical engineer in these areas of course work: chemical engineering, chemistry, mathematics, physics, humanities, and social science.

The case study will be devided into the following areas:

Background

A description of Rob, his educational background, his work experience, and his responsibilities.

Methodology

An examination of the method used to conduct the interview and collect information during the interview.

Views

A report on Rob's answers to each of my questions and my analysis of his comments.

Analysis

An overall analysis of Rob's comments

Portriat of a Chemical Engineer

A summary of Rob's answers that will give the reader a sense of Rob's views.

Advice

Proposed suggestions, based on Rob's advice, to improve the chemical engineering curriculum.

Directions for Further Research

Additional methods and possibilities for further research that might provide pertinent information about my concept of interst.

Preparation: Chemical Engineering Courses

1. Do you believe that you were sufficiently prepared academically to enter the workplace through chemical engineering course?

Answer

Rob responded to this question very positively. He believed that he was more than sufficiently prepared to enter the workplace as a chemical engineer. He began working at Xionide with four other engineers, and he found he was the best prepared of them all. He believed this was true because his department had trained him in state-of-the-art, market driven concerns, while maintaining the essential principles of chemical engineering. "They [the professors] knew what they wanted and shaped the curriculum accordingly," he said.

Analysis

Sufficient preparation is necessary for success in the chemical engineering work-place. In his university's curriculum, Rob found a sense of direction and focus that later paid off. This sense of direction agrees with an AIChE study on the skills needed for a chemical engineer. As the authors on the New Technology Committee write, "One problem requiring ongoing review is how to best educate chemical engineers to meet future needs. . . . a chemical engineering department today might select three major thrusts for its program (e.g., energy, environment, and advanced materials). The choices would be based on an assessment of market needs, competition from other disciplines, availability for financial support, relationship to other strengths in the university, and current department strength" (1987, p. 8). It seems that Rob's curriculum was focused in the way that the report suggests.

Reference Section
(a selection)

References

Floriman, S.C. (1984, Summer). The engineering view. *The Bent of Tau Beta Phi,*, pp. 26-32.

Layton, E.T. (1986). *The revolt of the engineers: Social responsibility and the American engineering profession.* Baltimore: Johns Hopkins University Press.

New Technology Committee of the American Institute of Chemical Engineers. (1987). *Chemical engineering: What skills are needed?* Washington, DC: AIChE.

Documentation System
The American Psychological Association
(Adapted for Undergraduate Student Use)

Journal Articles

Markusen, A. (1986). The militarized economy. *World Policy Journal, 3*, 495-516.

Books

Benton, L, Bailey, T. R., Noyelle, T., & Stanback, T. M., Jr. (1991). *Employee training and U.S. competitiveness: Lessons for the 1990s.* Boulder: Westview Press.

Chapters in Edited Volumes

Allen, W. (1979). The service sector economy in Newark. In Stanley B. Winters (Ed.), *From riot to recovery: Newark after ten years* (pp. 159-172). Washington, DC: University Press of America

Materials Presented Orally

Franck, K. A. (1989, October 22). *Routing, ritual, and the design of everyday life: A course.* Paper presented at the meeting of the Conference on Built Form and Culture, University of Arizona.

For more information about the APA system of documentation, see the following:

American Psychological Association. (1984). *Publication manual of the American Psychological Association.* (3rd ed.). Washington, DC: Author.

Assessment Criteria for the Field Research Paper

Below are listed the criteria for an effective *field research paper*. After reading the exhibit, please circle the response indicating your judgment as to whether the student has exhibited

4 - Superior ability

3 - Competent ability

2 - Somewhat competent ability

1 - Lack of competent ability

1. The writer provides a well focused aim (i.e., concept of interest) for the study. The organizational pattern for the study is established.

4   3   2   1

2. The writer clearly establishes the methods used to conduct the interview. The methods are clearly related to the aim.

4   3   2   1

3. The writer is quite able to analyze coherently the engineer's responses to questions. The analysis is specific, and published research is used to supplement observations.

4   3   2   1

4. A page of references is provided and rendered according to a recognized documentation system.

4   3   2   1

Discussion Questions for the Field Research Paper

1. Does the field study given in the exhibit appear to have a well-focused, precise research interest?

2. Do you think that the concept of interest is well-elaborated? Is it too broad or too narrow?

3. Does the writer provided a lucid summary of the subject's answer to the question about the preparation provided by chemical engineering courses?

4. Does the writer use secondary sources very well to further analyze the subject's answers?

---

[1] For further information about preforming field studies, see the following sources: Stephen Dohney-Farina and Lee Odell, "Ethnographic Research on Writing: Assumptions and Methodology," *Writing in Nonacademic Settings*, Eds. Lee Odell and Dixie Goswami, New York: Guilford, 1985; Norbert Elliot and Paul Zelhart, "Hermeneutics and the Teaching of Technical Writing," *The Technical Writing Teacher*, 17 (1990), pp. 150-164; Jack Selzer, "The Composing Process of an Engineer, "*College Composition and Communication*, 34 (1983), pp. 178-187; and Dorothy A. Winston, "Engineering Writing/Writing Engineering," *College Composition and Communication*, 41 (1991), pp. 58-70.

The Historical Research Paper

Introduction

It is very fulfilling to bring a well formulated idea together in a single document. The research paper is best thought of as the culmination of an effort that has included, at first, your own ideas and, then later, the research of others.

Traditionally found in history courses, the research paper allows you an opportunity to discover what others before you have thought and experienced. In the case of chemical engineering, such historical awareness is essential; without it, you cannot be a compleat engineer; with it, you will be able to put developments in your field into perspective.

Organization of the Research Paper

There are many ways to organize an historical paper. The following outline is one way that might be helpful to you.

Introduction

Provide a through, comprehensive introduction. Note, as the exhibit demonstrates, that this introduction leads the reader into the essence of the paper.

Background

Provide enough focused information so that the reader is given a context in which to put the information you will provide in your paper.

Analysis

Generate the analysis and explore the details of the analysis. Be sure to maintain the narrow focus of your paper as you go. Remember, you are presenting an argument in this paper, so details that do not contribute to your focus should be omitted.

Conclusion

Come to something about the topic. At the end of her paper, make some meaning of your historical analysis.

References

Provide a set of references. In the exhibit, we provide a documentation format used by the Modern Language Association.

Description of the Exhibit

In this paper the student writes about the beginning of the field of chemical engineering. The student attempts to relate the origins of the field to national trends in technology development at the turn of the 20th century.

Origins Of Chemical Engineering (1870-1920)

For the field of chemical engineering, it may be argued that the most important technological advances are not those which have occurred in the most recent past. Perhaps the key years are those at the turn of the twentieth century. The technological advances which occurred then are the advances which have shaped and formed the field of chemical engineering into what it has become today.

Basically, the advances of the period between 1870 and 1920 are based on the transition of centuries-old craft traditions grafted onto science (Layton 562). Even more importantly these fifty years have brought into being a nascent type of engineering that no one had believed was a true field in itself. That field is chemical engineering, a field which in the past was thought to be just a sub-field of mechanical engineering and chemistry (Marcus and Segal 172). However, before we look at the formation of the field of chemical engineering, let us first take a look at some other technological advances that took place during this period of history.

The Beginning of the Modern Period and the Birth of Engineering Organizations

Americans in the years between the 1870s and the 1920s proved far more discriminating than their predecessors. Americans living during this time tried in one way or another to put all things including people and ideas into some sort of category; they were obsessed with identifying diversity, treating each category as real, crucial, and limiting (Marcus and Segal 139). That is why this period in America's past is called by these historians as the "systematizing of America." However, putting everything that exists in this world into a category is very discriminating and detrimental to American society. This is especially true, when people are put into these categories and separated because of them. However, one good thing that did come out of the systematizing of America is that during this time period the mania translated into an increase in industry output and an increase in the flow of capital (Marcus and Segal 140) which, in turn, meant a boom in the economy.

This boom in the economy also led to most of the technological advances of this time, the result of efforts to rationalize and categorize the inventive process which prompted the establishment of research laboratories and served as guide-lines on how to invent. Some of the technologies that have benefitted from this process of thinking and systematizing are Edison's electric light, electric power, the telephone, and the field of engineering. Specifically, the field of engineering benefitted in that it was categorized into specific fields such as civil and chemical engineering. Consequently, each field was able to justify itself as autonomous.

These engineering specialties began about 1870 when the first engineering organization, the American Society of Civil Engineers, established itself to organize America's engineers (Marcus and Segal 165). However, because there are many fields of engineering, this one society could not compensate all of the fields. In turn more specialized organizations were formed including the American Institute of Mining Engineers (1871), American Society of Mechanical Engineers (1880), American Institute of Electrical Engineers (1884), American Institute of Chemical Engineers (1908), and finally the Institute of Radio Engineers (1912) just to name a few (Marcus and Segal 165). This organization of specific fields of engineering is just the beginning of engineering advancements; ultimately it was the main factor in getting the ball rolling for the further advancements of all engineers. By putting engineers into a different category it had separated engineers from other people in that they were no longer mere producers of products, they were now professionals with a professional code of conduct.

The engineering organizations also systematized the technical education of future engineers to further engineering's foothold as a profession in modern society. Original technical engineering education began because of military needs for professions to build weapons and fortifications. The first engineering college was formed at West Point in 1802, but because of a more industrialized society many more private colleges were shortly formed to keep up with the modern technology (Armytage). This increase in colleges led to a diversified education of engineers. By systematizing future engineers' educations the organizations would be sure that all engineers in each specific field would have the same knowledge of engineering, which could now be backed with a degree. This in turn would limit discrepancies in the public mind as to whether or not engineers as a whole could be considered professionals. Organizations also set the basic curriculum that would be the backbone or basic knowledge of all engineers; this knowledge would be based in the study of mathematics and physical science. This knowledge separated the engineer from the mere laborer.

Another problem the organizations had settled was whether the engineer should be more concerned with theory or practice. On one hand, it was said that the engineer should learn theory, and practice on the job. The reasoning was that theory separated the engineer from the mere laborer. However, mere theory without any practical experience rendered the engineer useless. In 1906 Herbert Schneider's cooperative education program at the University of Cincinnati attempted to harmonize theory and practice. Under his program, the young engineer spent half his time learning theory and the other half working in industry. By following this method the young engineer experienced the full spectrum of engineering experiences before entering the field (Marcus and Segal 171).

Our Field: Chemical Engineering

Now we will get more specific in that we will now discuss the actual creation and advancement of the engineering field known as chemical engineering. The field of chemical engineering went through the same advancements as previously stated;

however, as explained in the introduction, chemical engineers had problems defining themselves as a specific field and not a sub-field of mechanical engineering and chemistry. Because of these problems chemical engineering did not organize itself until 1908 with the creation of the American Institute of Chemical Engineers (AIChE). However, not until World War I did chemical engineering show itself to be fully autonomous. It was A.D. Little who showed to the public in 1915 the uniqueness of the field of chemical engineering in his letter to the president of MIT (Reynolds 10). A.D. Little did this by coining the phrase of "unit operations" in his letter which he felt constituted the building blocks of industrial chemical processes. Because of A.D. Little, the entire education of chemical engineers was revamped and reorganized, with the help of the AIChE, to fit this idea of unit operations into the chemical engineer's education. Unit operations consisted of the many processes used by chemical engineers to take certain reactants and to form and purify the desired products using a rational and economically-sound system consisting of these operations. Some basic unit operations are pulverizing, dyeing, roasting, crystallizing, filtering, evaporation, electrolyzing (Marcus and Segal 172). The AIChE quickly recognized the idea of unit operations and Little was rewarded by the organization. Because of the concept of unit operations proposed by Little and the help of the AIChE the field of chemical engineering was spearheaded into what we know chemical engineering to be today.

Further investigation by the AIChE also showed the importance of the laboratory research as a hands-on form of industrial practice and has shaped what we understand to be a chemical engineering lab today. During this time most labs were simply used to show students only the basic established scientific principles and not to establish new knowledge or new principles. Not until the 1870s did the idea of the research laboratory really come into play (Hounshell and Smith). It was Johns Hopkins University that started the idea of the research laboratory. By introducing this idea of the research laboratory, the American colleges had shaped industry itself. Industry had formed its own research facilities to better streamline existing factories through the knowledge gained from the research done in lab experimentation. The first modern research facility that used this method of experimentation was Du Pont's high explosives laboratory (Sturchio 10). So, because of research there is an increase in scientific capital (Layton 563) that snowballs into great scientific discoveries.

In conclusion, I hope it has been shown that the period of time from the 1870s through to the 1920s was the birth of engineering and technology as we know it to be today. Also, through the help of the American Institute of Chemical Engineers (AIChE) and especially through the help of A.D. Little, who could be defined as the father of chemical engineering, it has been shown that this period of time is what had defined chemical engineering into the specific field of engineering that we know it to be today.

Works Cited

Armytage, W. G. H. *A Social History of Engineering*. New York: Pittman Publishing, 1961.

Hounshell, David A. and John K. Smith, Jr. *Science and Corporate Strategy: Du Pont R&D, 1902-1980*. New York: Cambridge UP, 1988.

Layton, Edwin. "Mirror-Image Twins: The Communities of Science and Technology in 19th-Century America." *Technology and Culture* 1971: 562-   680.

Marcus, Alan I. and Howard P. Segal. *Technology in America: A Brief History*. New York: Harcourt, 1989.

Reynolds, Terry S. *Seventy Five Years of Progress: A History of the American Institute of Chemical Engineers*. New York: AIChE, 1983.

Sturchio, Jeffrey L. "Chemistry and Corporate Strategy at DuPont." *Research Management* Jan.-Feb. (1984): 10-18.

Documentation System
The Modern Language Association
(Adapted for Undergraduate Student Use)

Journal Articles

Katz, Eric. "The Call of the Wild: The Struggle against Domination and the Technological Fix of Nature." *Environmental Ethics* 14 (1992): 265-273.

Books

Opie, John. *The Law of the Land: 200 Years of American Farmland Policy*. Lincoln: U of Nebraska P, 1987.

Chapters in Edited Volumes

Heath, Shirley Brice. "The Fourth Vision: Literate Language at Work." *The Right to Literacy*. Eds. Andrea A. Lunsford, Helen Moglen, and James Slevin. New York: Modern Language Association, 1990. 289-306.

Materials Presented Orally

Katz, Eric. "Towards an Ethic of Partnership." United Nations. Department of Public Information and Non-Government Organizations Conference, "Environment and Development: Only One Earth." New York, 14 Sept. 1989.

For more information about the MLA system of documentation, see the following:

Gibaldi, Joseph and Walter S. Achtert. *MLA Handbook for Writers of Research Papers*. 2nd ed. New York: Modern Language Association, 1988.

Assessment Criteria for the Historical Research Paper

Below are listed the criteria for an effective *research paper*. After reading the exhibit, please circle the response indicating your judgment as to whether the student has exhibited

4 - Superior ability

3 - Competent ability

2 - Somewhat competent ability

1 - Lack of competent ability

1. The first paragraph of the paper is concise, focused, and lucid.

        4   3   2   1

2. An intriguing lead section has been provided that entices the reader to stay with the writer throughout the paper.

        4   3   2   1

3. In a sentence or two, the core idea of the paper is presented.

        4   3   2   1

4. Ideas have been elaborated so that each stands out distinctly from the others

        4   3   2   1

5. The sum of these ideas provides a developed argument for the thrust of the paper?

        4   3   2   1

6. A conclusion is provided that fulfills the expectations established in the beginning of the paper? (Hint: Read the sentences you identified above in Question 3 against the conclusion of the paper? Are these sections complementary?)

        4   3   2   1

7. The writer has followed an established documentation system. There are no documentation errors that proved distracting to you, thus weakening the overall effect of the paper?

        4   3   2   1

Questions for Discussion of the Research Paper

1. Consider the overall design of the paper. Do you believe that the writer has structured the paper convincingly?

2. Clearly, a major factor in a research paper is its use of sources. Locate two uses of sources and disucss why you find them especially effective?

3. What else would you add to the paper to make it richer?

4. When you are given the opportunity to select a topic for a history course, do you take this as an opportunity to study the origin and development of your field? If so, what kinds of papers have you written? If not, why don't you select such topics?

11 Forms of Communication for Students of Chemical Engineering: In the Laboratory

Innovation, complexity, intricacy, social influence, and simple extensiveness of the corpus make scientific writing interesting as an object of study and important as a part of human society.

Charles Bazerman, *Shaping Written Knowledge: The Genre and Activity of the Experimental Article in Science*

A Look at Communication in the Chemical Engineering Laboratory: National Trends

As we were preparing this textbook, we reviewed chemical engineering laboratory courses in many of our nation's leading universities. What we suspected was confirmed: the laboratory is a center for instruction in critical thinking and a forum for technical communication.

Here are some examples from universities across the country:

Clarkson University (Potsdam, New York). In the unit operations course, students communicate with their instructor in a three-stage procedure. First, students submit a preliminary written report the day prior to the scheduled laboratory period. Written as technical memoranda, these reports include the following sections: a memo of transmittal; a table of contents; an introduction (1–2 pages); a background and theory section (4 page maximum); a description of the equipment (2 page maximum); an experimental plan (3 page maximum); a data analysis section (3 page maximum); a discussion (1 page maximum); and a list of references. This preliminary report must be judged to be acceptable by the instructor before the laboratory experiment is performed. In stage two, an oral progress report is given by the group. Here, sample calculations, graphic illustrations and/or tabulations are presented for all results obtained during the first laboratory period. As well, an outline is presented of the procedure the group will follow during the second laboratory period. In the third phase, students submit a final written report upon completion of the experiment. The final report includes these divisions: a memo of transmittal; a summary (1 page maximum), a table of contents; an introduction (2 page maximum); an experimental methods section (2 page maximum); an experimental results section (4 page maximum); a discussion section (4 page maximum); a conclusion (1 page maximum); a list of literature references; and appendices, including detailed calculations, raw data, graphs and tables of intermediate results, and any additional information. Data Books are kept throughout the experimental process.[1]

Colorado School of Mines (Golden, Colorado). Students in the unit operations laboratory begin by preparing a detailed experimental plan submitted in the group leader's notebook. The notebook must provide the following information: a concise statement of the experimental objectives; a discussion of the theory; working equations; a detailed plan of operation, including a sketch of the apparatus and a list of all necessary equipment and supplies; a set of data tables for recording experimental results; a discussion of safety precautions; and a list of duties for each group member. When the supervising faculty member is satisfied with the group's experimental plan, a grade is assigned to the notebook and the group begins the experiment. When the experiment is completed, students submit either an informal short written report, an oral report, or written final report. In addition, a member of the Humanities and Social Science Department who is a specialist in technical communications works closely with the students.[2]

University of Arkansas (Fayetteville, Arkansas). Upon enrolling in CHEG 1212 (Chemical Engineering Laboratory I, Experiments to Determine Physical Properties of Importance in Engineering) or CHEG 3323 (Chemical Engineering Laboratory II, Experiments in Momentum and Heat Transport), students are told that a report "must not only be timely, it must be technically sound and of good quality." While there are guides for grammatical correctness, "only practice, attention to the needs of the reader, and careful editing of what you have written are necessary if [you] are to write correctly as well as efficiently." To convey the results of their experiment, students compose long reports, short reports, and memoranda.[3]

University of California, Berkeley (Berkeley, California). Students in the chemical engineering laboratory must keep a detailed project notebook. "What you write," they are reminded, "should be complete, orderly, and understandable by others (as well as yourself when reading it a few years later)." For less-complex experiments (e.g., fluid flow measurement), students present a fifteen-minute oral report on the results of the experiment. For more complex experiments (e.g., esterification kinetics), students begin with a planning conference in which they present orally their experimental design. As the experiment proceeds, students also must orally present their work at briefing and design conferences. Final work is submitted in a formal written report.[4]

The University of Washington (Seattle, Washington). Students who enroll in ChE 436, a transport laboratory, find themselves in the role of employees of Seattle Chemical Industries.[5] Assignments such as the following are given in memo formats:

Seattle Chemical Industries
Engineering Development Laboratory
Seattle, Washington 98195

April 1991

To: Group Leader
From: Profs. C.A. Sleicher, E. M. Stuve
Subject: Evaluation of Friction Losses and Pipe Fittings

The Engineering Construction Section of SCI must frequently estimate pumping requirements for flow systems which they have designed. They have asked us to determine whether the generalized friction loss information in the literature is adequate for the specific pipe and tubing that we typically use. For this purpose, we have had a flow loop constructed in our laboratory with pressure taps installed on each branch. We have three different diameters of rough pipe and have matched one of those in smooth tubing. In addition, we have a station in which various fittings may be installed for evaluation.

We would like you to establish a friction factor-Reynolds number relationship that can be used for design purposes and assess the errors therein. Please also evaluate the losses to be expected in typical fittings. Try to be as general as you can and cover the widest Reynolds number range possible with this equipment.

We would like your reports on this project at our section meeting three weeks from today.

The University of Dayton (Dayton, Ohio). A capstone approach has been used in the unit operations laboratory at this university. A one-statement assignment begins the course: "Improve the quality of our department laboratories by designing new experiments." Students begin the project by evaluating the status of laboratory experiments used in four different laboratory courses at the university. They establish goals for each of the four laboratories and specify whether the experiments meet these goals. To refine these goals, students contact various chemical engineering departments to obtain information about other laboratories. This first phase of the course is concluded when students generate criteria that will be used to evaluate and guide the experimental process that they will design. In phase two of the course, student begin designing their own experiments around the apparatus they critically examined in the first phase of the course. In the unit operations laboratory, for example, students decided to design systems for reverse osmosis purification of water, flash distillation of a liquid mixture, tubular chemical reactor with photometric analysis, and a semibatch chemical reactor. To communicate their collaborative work in both parts of the course, students submit their work in a variety of oral and written presentations.[6]

In our review of these and other programs, it is clear that departments believe, as Ralda M. Sullivan writes in her review of the program at Berkeley, that "there is a direct correlation between professional advancement and the ability to write and speak effectively."[7] We would add that this ability is a manifestation of critical thinking. Because departments acknowledge this relationship, there is a great deal of effort spent in developing innovative ways to demystify communication for students. In this chapter we present concrete examples of various ways to report your laboratory work. Our effort is to help you realize that effective communication—the ability to refine and capture your critical thinking—is within your grasp.

Simulation of the Organizational Reader/Writer Relationship

In Chapter 10, the papers we presented were frequently written for an isolated academic audience: the instructor. But, in the workplace your communications will be written for multiple readers operating in complex organizational structures. How can we re-create these more complicated writer/audience relationships within the undergraduate chemical engineering laboratory?

On one level, we cannot, nor can anyone teaching in a university. The student's chemical engineering laboratory takes place within an academic setting, a very real world of credit hours and grade point averages. However, we can create a simulation by means of various reporting structures addressed to various types of readers. These simulations will allow you to anticipate and thus master the diverse communications situations you will encounter later in the workplace.

What You'll Find in the Discussion of Each Report

In the following pages we will present guides that will enable you to produce nine reporting structures:

- the laboratory notebook,

- the briefing memo,

- the technical translation memo,

- the scale-up memo,

- the formal laboratory report,

- the procedural report,

- the proposal request for funding,

- the scholarly paper, and

- the oral presentation proposal.

As is the case with each of the types of reporting structures you studied in the previous chapter, in our presentation of each form of writing from the laboratory you will find:

- a brief *introduction* to the reporting structure,

- a section on *organization* to guide you in preparing the report

- a brief *description* of the experiment on which the report is based,

- a sample of the competed report, called the *exhibit*,

- a series of *assessment criteria* by which the report may be judged, and

- *questions for discussion* about the exhibit which generally refer to the annotated instructor's comments.

Obviously, these descriptions and exhibits are not to be taken as cookbook exercises. If you are to benefit by using them, you should discuss them with your classmates and your instructor to determine which strategies seem to work best.

A Note on the Creation of the Exhibits

The exhibits in Chapter 11 were created by our students in a two-semester senior chemical engineering laboratory. Working in groups, students used the critical thinking model upon which this textbook rests. All papers were written under the writing process model provided in Chapter 9. All papers were drafted by our students, revised under our supervision, and then re-submitted for a final grade.

As is the case in Chapter 10, we have selected what we believe to be solid sample student papers which successfully fulfilled the experimental assignments and aptly illustrate a rich array of critical thinking abilities.

We have selected samples of student work—rather than created our own perfect samples—for three reasons. First, it is our experience that students are quite interested in what other students have done at their same level of experience and ability. Second, students are interested in the comments that instructors make on student work in that these comments articulate the kind of tacit criteria the instructors use to grade the papers. Third, we have used student samples without totally correcting them because perfect samples, written by instructors and copy-edited by publication professionals, often prove daunting to students. While the exhibits may be somewhat flawed, they are real and thus represent an obtainable goal that you may strive to fulfill and surpass. In a sense, the exhibits are the results of our experiments with the approaches we have advocated throughout this book. The validity of our approach must stand or fall on the merits of these examples.

For the sake of brevity, we show in many of the exhibits only those portions which render the report unique. For example, while an Appendix consisting of supporting data and sample calculations is called for in many of the structures, these have not been shown.

You will also notice that we have commented on the exhibits. We believe these annotations will be useful in that they will provide you with an idea of the kinds of questions your laboratory instructor might raise about your work. These annotations also simulate the kinds of comments that you will receive on your work both within and beyond the university.

In essence, we hope that you will use these interactive exhibits to promote discussion on the communications strategies that will enable you to become a compleat chemical engineer.

Notes and References

1. Material on Clarkson University's program was provided by Richard J. McCluskey, Associate Professor and Executive Officer.

2. Material on the Colorado School of Mines program was provided by Robert M. Baldwin, Professor and Head.

3. Material on the University of Arkansas' program was provided by William M. Myers, Assistant Department Head.

4. Material on the program at the University of California at Berkeley was provided by Fred Vorhis, Lecturer and Lab Coordinator.

5. Material on the University of Washington's program was provided by Bruce A. Finlayson, Rehnberg Professor and Chairman.

6. Material on the University of Dayton's program was provided by Kevin J. Myers, Assistant Professor. The capstone project has been presented by Dr. Myers in the *Proceedings of the North Central section of the American Society for Education Conference* (Sanginaw Valley State University, April 4–6, 1991).

7. Ralda M. Sullivan, "Teaching Technical Communication to Undergraduates: A Matter of Chemical Engineering," *Chemical Engineering Education*, Winter, 1986, pp. 32–35. We note in passing that the emphasis on communication in chemical engineering extends beyond the laboratory. See, for example, the writing emphasis in Bucknell University's plant process design course described in *Design the Write Way: Writing in Engineering Design Courses*, Michigan Technological University, Issue # 3, Spring 1991, pp. 5–6.

The Laboratory Notebook

Introduction

Congressman John Dingell had the Secret Service run forensics tests on the laboratory notebooks of Tereza Imanishi-Kari. The findings, released in 1989, revealed that Dr. Imanishi-Kari had falsified the data for her publication in *Cell* on gene immunology. This discovery of falsified data shook the scientific community: a co-author of the paper was Nobel-Prize-winner David Baltimore. [1]

The episode points out the importance of keeping lucid, truthful records of your laboratory research.

Why pay attention to the laboratory notebook? Senior Research Chemist Howard M. Kanare provides these reasons:

- The notebook preserves experimental data and observations that are part of the experimental process itself.

- The details provided assure that another researcher could pick up the notebook, repeat the experiment, and make similar observations. This process of replication is fundamental to the scientific process.

- The notebook is the first step in recording the data and making the observations that eventually lead to writing reports, technical papers, patent disclosures, and correspondence with colleagues. [2]

The laboratory notebook may therefore be considered as the place where critical thinking begins to be articulated.

Organization of the Laboratory Notebook

Kanare recommends that the laboratory notebook be organized as follows:

Organization of the Laboratory Notebook

Front Matter

An *exterior title* should be provided on both the spine and front cover. Titles should be kept brief.

A *printed signout page* should record the date of issuance of the notebook.

A *table of contents*, similar to that provided in Exhibit 1, should provide dates, subject, and page numbers.

A *preface* should describe the goal of the research that will be recorded in the notebook, the location of the research, and, if applicable, the source of research funding.

A *table of abbreviations* should provide codes for information found in the notebook

229

Each experiment should have the following sections:

Introduction: to state the purpose of the research

Experimental plan: to set out the path of your research; the direction of your experiment should be accomompanied by a flowchart

Observations: to record your data

Conclusions: to illustrate the relationship between the purpose of the research and the observations; this section should include a statement of further considerations involving unanswered research questions

As the Baltimore case illustrates, there are both legal and ethical issues involved in the laboratory notebook. In addition, it is the first place for you to use writing as a way of learning. "Thinking," as philosopher Martin Heidegger said, "is hand work." There is no better place to begin this work than with the laboratory notebook.

Description of the Experiment

The exhibit is based on an experiment in yeast growth. Here, the reaction kinetics of yeast growth and glucose consumption are studied. The experimental objectives include the following: determination of cell mass and glucose concentrations as functions of time, and 2) application of chemical reaction engineering and modeling principles to a biological system.

003

FERMENTATION   1/28

KINETICS OF YEAST GROWTH

OBJECTIVE: TO DETERMINE THE KINETICS OF YEAST
GROWTH BY STUDYING ITS BIOLOGICAL
PROCESS.

BACKGROUND: THE KINETICS OF YEAST GROWTH WILL
BE DETERMINED USING A FERMENTOR —
DISTILLED WATER, GLUCOSE, NITROGEN
BASE, + YEAST ARE ADDED TO THE
FERMENTOR — AIR FLOWS THROUGH THE
FERMENTOR WHILE pH + TEMPERTURE
REMAIN CONSTANT — SAMPLES ARE THEN TAKEN
PERIODICALLY TO DETERMINE GLUCOSE +
YEAST CONCENTRATIONS. SUCH CONCENTRATIONS
WILL BE USED TO AID IN THE DETERMINATION
OF THE KINETICS BY THE FOLLOWING
REACTION:

YEAST + SUGAR + OTHER   --AIR--> YEAST
        (GLUCOUS) (N_2 etc.)

APPARATUS: 1. FERMENTOR

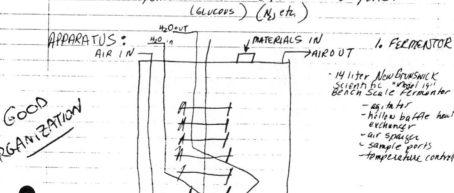

- 14 liter New Brunswick
Scientific "Model 19"
Bench Scale Fermentor w
- agitator
- hollow baffle heat
exchanger
- air sparger
- sample ports
- temperature control

GOOD
ORGANIZATION

231

007

FERMENTATION   2/4

KINETICS OF YEAST GROWTH

Add Yeast 9:20 Am , ✻ = Added NaOH, ● dilution 10/1,
X Dilution 5/1

TIME	YEAST CONC (ml/ml)	pH		Absorbance		Glucose Conc mg/ml
9:20	.075/14	✓		.69	●	33.2
9:35	.075/14	✓		.59	●	33.2
10:00	.1/14	✓		.585	●	32.7
10:30	.05/14	✻		.570	●	30.31
11:00	.06/14	✓		.46	●	27.5
11:25	.06/14	✓		.45	●	25.2
11:40	.1/14	✓		.43	●	23.3
11:55	.12/14	✓		.38	●	20.73
12:10	.13/14	✓		.28	●	19.5
12:25	.15/14	✓		.18	●	18.85
12:40	.17/14	✓		1.9	X	17.46
12:55	.17/14	✓		1.3	X	16.3
1:10	.17/14	✓		.83	X	15.2
1:25	.18/14	✓		.95	X	13
1:50	.12/14	✓		.75	X	11.9
2:10	.12/14	✓		.65	X	10.0
2:30	.12/14	✓		.5	X	10.0

TEMP TOO HIGH 40°C

WELL ORGANIZED & NEAT

CONCLUSION:

YEAST CONC INCREASED → LEVELED OFF
GLUC. CONC STEADILY DECREASED

A QUICK PLOT WOULD BE USEFUL HERE IN NOTEBOOK

TEMPERATURE GOT EXCESSIVELY HIGH SPEEDED UP YEAST GROWTH! — WHERE? ABOVE

232

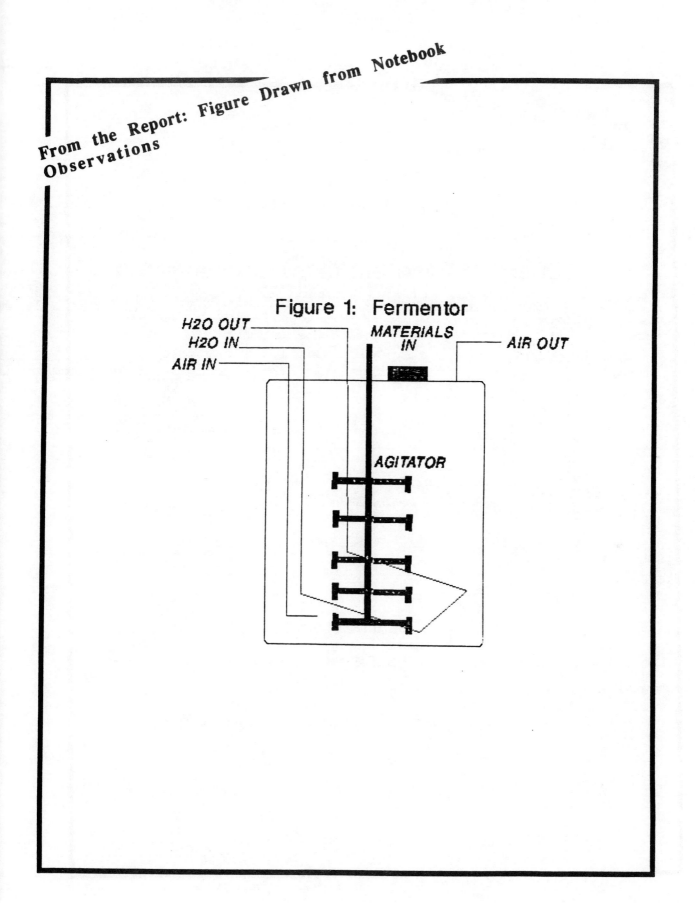

Figure 1: Fermentor

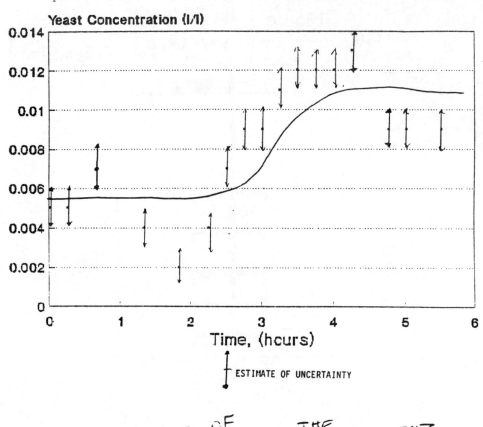

Effect of Time on Yeast Concentration

Yeast Concentration (l/l)

ESTIMATE OF UNCERTAINTY

GOOD USE OF UNCERTAINTY BARS — THE DIFFICULTY OF THE MEASUREMENTS ARE CLEAR

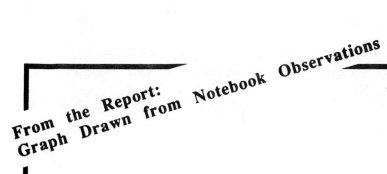
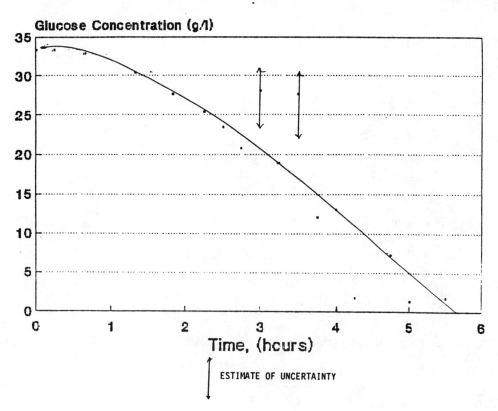

Effect of Time on Glucose Concentration

Glucose Concentration (g/l)

Time, (hours)

ESTIMATE OF UNCERTAINTY

Assessment Criteria for the Laboratory Notebook

Below are listed the criteria for a well kept *laboratory notebook*. After reading the exhibit, please circle the response indicating your judgment as to whether the group has exhibited

4 - Superior ability

3 - Competent ability

2 - Somewhat competent ability

1 - Lack of competent ability

1. An exterior title is provided on both the spine and front cover. The title is brief.

4   3   2   1

2. A printed signout page records the date of issuance.

4   3   2   1

3. A table of contents provides dates, subject, and page numbers.

4   3   2   1

4. A preface describes the goal and location of the research.

4   3   2   1

5. A table of abbreviations provides codes for the information found in the notebook.

4   3   2   1

6. An introduction states the purpose of the research.

4   3   2   1

7. An experimental plan sets out the path of the research. A flowchart is included.

4   3   2   1

8. Well written, clear, and concise observations are made in reference to the data.

4   3   2   1

9. Conclusions illustrate the relationship between the purpose of the research and the observations.

4   3   2   1

10. There is a statement of further considerations involving unanswered research questions.

4   3   2   1

Discussion Questions on the Laboratory Notebook

1. How is the exhibit an example of the interdependence of writing and research discussed in Chapter 9?

2. Why are headings and sub-headings so important to the organization and clarity of laboratory notes? Illustrate your answer by reference to the exhibit.

3. Why is a simple schemata, such as that found in the notebook, so important?

4. How did the table in the laboratory notebook enable the students to make the graphic comparison of (1) the effect of time on yeast concentration and (2) the effect of time on glucose concentration?

5. Identify instances of critical thinking in both the laboratory notebook and the finished graphs.

---

[1] For more on the Baltimore case, see O. Weiss, "Conduct Unbecoming," *The New York Times Magazine*, (October 29, 1989), pp. 41-71, 95.

[2] Howard M. Kanare, *Writing the Laboratory Notebook*, Washington, DC: American Chemical Society, 1985.

The Briefing Memo

In 1985 Paul V. Anderson analyzed fifty of the best modern surveys on writing at work. Anderson found that the kind of documents most commonly written in the workplace are letters and memoranda.[1]

Why are memoranda so popular? Two reasons may be given:

- Memos are brief and therefore may be read quickly by busy people in the organization

- Memos are used to inform others in the organization about work in progress.

However, because of their brief nature, many memos are neither carefully structured nor articulate. "It's only a memo," is the common excuse. Remember, though, that in an organization the only impression someone may get of you is through your writing; a hastily written document may create an impression that will be hard to erase. Remember, too, that there is no more important act than briefing others on your work; if you submit a shabby account of your work-in-progress, then questions may arise about the worth of your research.

What are the positive benefits of producing crisp, coherent memos?

- You can inform others of your work and benefit by their comments.

- You can combine these shorter documents into required longer documents.

- You can create an identifiable voice in your organization, one that stands for intelligent thought and high standards.

Recall the thesis of this book: you cannot be an excellent engineer without excellent communication abilities. Practice in creating masterful memos will be well rewarded.

Organization of the Briefing Memo

In an often-used form of laboratory communication, you will tailor the briefing memo to report on your current laboratory experiment. Sometimes, the memo is written to accompany other reporting structures; sometimes, the memo is written as a concise form of a more complex report. Regardless of its length or purpose, a memo can be structued in the following manner:

Organization of the Briefing Memo

Heading

Identify the receiver, the sender, the date, and the subject.

Make sure that the memo title, provided in the subject section, embodies the essence of the experiment.

Lead

Explain the background of the experiment, including chemical engineering theories and principles especially related to the experiment.

Procedure

Explain how the experiment was conducted, using a schematic diagram.

Results

Present your most significant findings, making sure that you distinguish between crucial and supplementary data.

Explain the differences between observed and experimental results.

Discuss the limits of your data.

Conclusions

Discuss the significance of your findings.

Provide a range of uncertainty.

Identify research problems that remain unanswered.

Make any recommendations on the experiment that you believe would improve the quality of the research.

Notice that the structure of this memo echoes the pattern of engineering research presented in Chapter 3. Thus, this brief document asks you to incorporate a formal laboratory report—and the critical thinking patterns of chemical engineering itself—into one or two pages.

The task is therefore very demanding. Experience tells us that a memo cannot be dashed off just a few minutes before it is due.

The Experiment

The exhibit is based on a continuous heat transfer experiment in which three heat exchangers are evaluated. The primary experimental objectives include the determination of selected overall and individual film heat transfer coefficients as a function of flow regime. In addition, for the same flow conditions, the performance characteristics of two of the exchangers are evaluated.

Introduction and Procedure

TO: DATE:

FROM: FILE:

SUBJECT: Characteristics of Shell and Tube and Flat Plate Heat
          Exchangers

CC:

NOTE : A SIMPLE SCHEMATIC WITH DATA COLLECTION SITES WOULD BE USEFUL

1.0 INTRODUCTION

The object of this investigation was to evaluate the characteristics of
a Basco size 03014 type 500 shell and tube heat exchanger, an American
Standard 02024 FCFB shell and tube heat exchanger, and an American Heat
flat plate heat exchanger by experimentally determining the overall
heat transfer coefficients (via heat balances) and individual surface
coefficients (via the Wilson plot method). The overall heat transfer
coefficient U_o is a measure of the overall resistance to heat transfer.
It is a function of resistances such as film build-ups due to dirt,
fluid films on the tube walls, and the conductivity of the tube walls.

GOOD CONCEPT

Dittus and Boelter upon studying the characteristics of heat exchangers
concluded the following correlation for liquids in turbulent flow
through tubes:

← WORDY

$$h_i D/k = 0.023 \ (N_{Re})^{0.8} \ (N_{Pr})^{0.33} \ (u/u_w)^{.14} \qquad \text{EQN 1}$$

$$Uo = \frac{(m_w)(Cpw)}{A_o} * \ln \frac{T_s - T_{in}}{T_s - T_{out}} \qquad \text{EQN 2}$$

WHAT ARE u ? AND u_w. SHOULD DEFINE

where: N_{Re} is characteristic of the flow, that is, the Reynolds is
larger for faster flow rates; N_{Pr} is characteristic of the heat
capacity, viscosity, and thermal conductivity of the liquid. It is
this correlation in which the experimental values of h_i are compared.

where: m_w=mass flow rate of cooling water; Cpw=heat capacity of
cooling water, A_o=surface area exposed to heat transfer; T_s=temperature
of entering steam; T_{in} and T_{out} are the entering and exiting liquid
temperatures respectively.

2.0 PROCEDURE

ROTAMETER DISCUSSION NOT IMPORTANT

The rotometer is calibrated by diverting the exit water flow to a tared
receiver and weighing over a period of time for the entire range of
inlet water flow rates. With cooling water flowing through the tubes,
steam is condensed on the shell side of a Basco Shell and tube heat
exchanger. The condensate is then forced into the tubes of an American

← JUST SAY "FLOWS"

240

Standard shell and tube exchanger and is cooled by water flowing on the shell side. The same procedure is repeated using the flat plate exchanger as the condenser (again with the American Standard as the post cooler). Figures 1-4 in the appendix are diagrams of the units used along with their characteristics.

OK

The temperatures of the inlet and outlet streams are recorded on a strip chart recorder and are used in energy balances to calculate the film and overall coefficients as well as heat gains and losses by the water and the steam.

GOOD – RELATE MEASUREMENTS TO HIGHER LEVEL RESULTS

3.0 RESULTS

Table 1 lists the values of the film coefficients as obtained from the correlation stated.

BASCO SHELL AND TUBE HEAT EXCHANGER

N_{Re}	h_o(emp)	h_i(emp)	U_o(emp)	h_o(exp)	h_i(exp)	U_o(exp)
19250	233.7	2363	196.5	247.3	4294	245.0
16300	247.1	2067	201.4	"	3992	238.4
14800	255.1	1916	203.9	"	3685	242.7
14000	259.5	1838	205.0	"	3371	229.4
10750	284.0	1481	209.3	"	2380	244.0
9260	298.4	1315	210.1	"	2027	231.4

GRAPHS WOULD BE MORE INSTRUCTIVE

GOOD RESULT

As seen from Table 1, a direct relationship exists between the flow (charcterized by the Reynolds number) and the inside film coefficient h_i. That is as the Reynolds is increased the film coefficient is increased. In fact this true for the American standard and flat plate exchangers as well.

For the American Standard shell and tube heat exchanger, h_i ranged from 103 to 186.4 Btu/hr·ft^2·^{o}F for N_{Re}: 487 to 1013. For the American Heat flat plate exchanger values of h_i ranged from 3200 to 4640 Btu/hr·ft^2^{o}F for N_{Re}: 8760 to 15500.

NOT NEEDED

4.0 CONCLUSIONS

- In every case, values of the inside film coefficient h_i agree with theory (increasing Reynolds numbers yield increasing values of h_i).

QUANTIFY

WHAT ABOUT UNCERTAINTY LIMITS?

- Individual values of the inside film, outside film, and overall heat transfer coefficients do not agree well with the values predicted by the empirical correlations.

WHY?

- At similar flow conditions, the overall heat transfer coefficient U_o for the American Heat flat plate condenser is 50 percent larger than that for the Basco shell and tube unit. Hence it can be concluded the the flat plate condenser facilitates a larger degree of heat transfer (per unit area and driving force).

GOOD! INSIGHTFUL

Assessment Criteria for the Briefing Memo

Below are listed the criteria for a *briefing memo*. After reading the exhibit, please circle the response indicating your judgment as to whether the group has exhibited

4 - Superior ability

3 - Competent ability

2 - Somewhat competent ability

1 - Lack of competent ability

The Briefing Memo

Heading

1. The receiver, sender, date, and subject have been identified. The memo title, provided in the subject section, embodies the essence of the experiment.

   4   3   2   1

Lead

2. The background of the experiment is explained, including chemical engineering theories and principles especially related to the experiment.

   4   3   2   1

Procedure

3. Explanation is made regarding how the experiment was conducted. A schematic diagram is included.

   4   3   2   1

Results

4. The most significant findings are presented. Crucial and supplementary data are distinguished.

   4   3   2   1

5. Differences between observed and experimental results are explained.

   4   3   2   1

Conclusions

6. The significance is discussed regarding the findings.

   4   3   2   1

7. A range of uncertainty is provided.

   4   3   2   1

8. Remaining unanswered research problems are identified.

4   3   2   1

9. Recommendations are made that would improve the quality of the research.

4   3   2   1

Discussion Questions on the Briefing Memo

1. How would a simple schematic and a single graph of the results have improved the document?

2. Using the audience analysis system provided in Chapter 9, what circle of expertise would you say the audience of the exhibit possesses? How can you infer the audience's level of technical understanding from this document?

3. Ask one of the instructors in your department to bring to class a superior memo written in a corporate setting and to distribute copies of the memo to the class. With your class, discuss how that memo is similar to and different from the exhibit in terms of the following: length, organizational pattern, level of technical information, and tone.

4. Why are uncertainty limits important? (Recall Chapter 4). How would the Results and Conclusions sections of the exhibit be affected if limits had been included?

---

[1] P. V. Anderson, "What Survey Research Tells Us about Writing At Work," in L. Odell and D. Goswami, eds. *Writing in Nonacademic Settings,* New York: Guilford Press, 1985. See also H. W. Swanson and H. M. R. Aboutorabi, "The Technical Memorandum: An Effective Way of Developing Technical Writing Skills," *Engineering Education,* May/June 1990.

The Technical Translation Memo

Introduction

Return to Chapter 9 and refer to Figure 9.5. You will recall that the chemical engineer reported that even those in his own group had very diverse levels of technical expertise. Again and again, both research and personal experience confirm that the ability to translate technical information to both specialists and non-specialists is an important critical thinking ability.

Why is technical translation important?

Technology transfer, a subject important to America's economic future, takes place primarily through written documents. Unless the source writer has mastered the ability to translate complex material to diverse audiences, there is little chance that technological concepts will be made accessible to target users.

The ability to translate knowledge is an important part in the process of understanding that knowledge. You have perhaps heard your instructors say that they never understood a given subject until they taught it. Why? Simply because teaching calls for translation of complex subjects to students unfamiliar with these subjects. In order to explain a difficult subject, you must master that subject. It is indeed useful to wonder whether deep understanding of a subject can occur unless an individual is able to translate the essence of that subject to others.

Translation is part of organizational life. The situation depicted in Figure 9.5 is not the exception. Frequently, you will have to report your research to those unfamiliar with your work. When funding decisions are made, your ability to convey the significance of your work becomes nearly as important as your ability to perform that work in the first place.

Translation is part of professional responsibility. A frequent criticism made against the scientific community is that researchers fail to communicate their work to others. Indeed, the stereotype of the absent-minded researcher probably prevails in the popular mind. It is important, then, for you to be able to explain your research to those who will be affected by its results.

Procedure for Writing the Technical Translation Memo

Three sequential steps should be taken when planning your technical translation.

Step 1. Decide the central concept of your subject and the amount of detail that will be necessary to explain that concept. A frequent problem with technical translation is that, in a effort to be thorough, a researcher will provide so much detail that the presentation becomes confusing.

Step 2. Use the audience analysis system discussed in Chapter 9. This technique will allow you to be sensitive to variance in your readers' level of understanding.

Step 3. Select appropriate translation strategies. We have found that four strategies are most effective:

Provide the historical background. The technique allows readers to contextualize the discussion that will follow.

Provide analogies. Explaining something unfamiliar in terms of something familiar allows your readers to become comfortable with the subject. When, for example, Albert Einstein wanted to explain the power of nuclear energy to an anxious world in 1946, he used the following analogy with a popular audience: Imagine that a rich miser, living in a small community, gives away no money (energy) during his lifetime. [1] When he dies, he stipulates in his will that one thousandth of his whole estate must be given to the community. So great is his wealth and so small is the community that the sudden influx of money (considered as kinetic energy) brings with it the threat of evil. After using the analogy, Einstein writes the following: "Averting that threat has become the most urgent problem of our time." Because Einstein anticipated that his audience would have trouble following a mathematical explanation of the law of the equivalence of mass and energy, he selected an analogy that conveyed his message: the world, posed on the brink of the nuclear age, had to be careful.

Provide visual representation. When you provide a simple schematic diagram or a block figure, you allow your audience to visualize your subject. Note the schematic in the exhibit.

Provide an illustration of the significance of the research. To return to Einstein's example, recall that he used his analogy to remind his audience that they needed to be aware of the threat of evil. Pointing out the importance and application of a topic is an excellent device for technical translation. Note that the lead paragraphs work so well in the exhibit because the use of fluid flow is discussed. [2]

We want to close our introduction with a warning. Never assume that your technical expertise allows you a privileged position. Our most vivid memory is of a student who interpreted our assignment to produce a technical translation as a call to write something that "any dummy could understand." The kind of interdisciplinarity that we have called for in this book demands that all fields of study conduct research in an atmosphere of mutual respect. It will do you no good to dismiss your group vice-president as unintelligent because he no longer works in the lab or to ignore a colleague's advice because his degree is twenty years old. Your desire to communicate, not your impulse to demonstrate your expertise at the expense of others, should motivate your translation of your work for others.

Organization of the Translation Memo

The structure we want you to follow for the translation memo appears below. A translation memo may—and perhaps should—accompany any of the reporting structures.

Organization of the Technical Translation Memo

Lead

Establish the aim and the significance of the research.

Apparatus

Provide a block diagram of the equipment used.

Your desire to communicate, not your impulse to demonstrate you expertise at the expense of others, should motivate your translation of your work for others.

Describe the procedure by which the equipment was operated, citing only the most significant steps.

Results

Provide the findings and explain why those findings are significant.

Conclusion

Provide an interpretation of the research

In each section of the memo, be sure to consider the amount of detail, central concepts, and use of vocabulary. Also be sure to use a variety of appropriate translation strategies throughout the memo.

Description of the Experiment

The exhibit is based on an experiment in which liquid flow through piping and fittings is studied. The primary objectives include: 1) the determination of the number of equivalent lengths and equivalent velocity heads lost by flow through various fittings, 2) the determination of the friction factor as a function of Reynolds number for straight pipe flow, 3) the evaluation of orifice and venturi meter flow coefficients, and 4) the evaluation of the characteristic curve for a centrifugal pump.

To: From:

Translation Memo for Fluid Flow Experiment

In most industrial processes materials must be
transported from one place to another. These materials are
usually fluids and must be pumped and stored throughout the
plant. It is therefore import to know the principles that
govern fluid flow, and the equipment that is used to
transport fluids.

A fluid is a substance that is in a liquid or gaseous
form. In order to transport a fluid in a pipe a driving
force is needed. This driving force is usually supplied by
pumps when transporting liquids and blowers when
transporting gases. The amount of fluid flowing in a pipe
can be measured using various meters. The amount of fluid
being transported can be calculated by the following
relation:

GREAT TRANSLATION OF A BASIC CONCEPT

(ALSO COMPRESSOR FOR GASES

Flow of material = driving force / resistance

Therefore in order to transport a fluid a pump must supply
enough pressure to overcome the resistance of the piping
system. In this case pressure is the driving force and the
resistance is the resistance of the straight length of pipe
plus the sum of all the resistances of the individual valves
and fittings. Therefore the total resistance of piping
system must be overcome in order to supply a given flow rate
through the pipe. By measuring the pressure drop across the
various valves and fittings the resistance of the valves and
fittings can be found. By measuring the pressure of the
inlet and outlet of a pump, characteristics of the pump can
be determined.

Apparatus and procedure:
The following is a simplified block diagram of the apparatus
used.

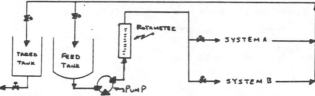

GOOD — JUST ENOUGH DETAIL

A pump was used to supply water from a feed tank to two
piping systems. The water then flowed back to the feed tank
or a measuring vessel. The flow rate of the water was

(handwritten) CALIBRATED → (ROTAMETER)

measured by the use of a flowmeter. The first piping system was two inch pipe containing two devices used for measuring the flow rate of the fluid going through the pipe. The second piping system which was 1/4 inch pipe contained several fittings and valves.

The procedure for the experiment was as follows:

1. Collect and weigh water for a period of time and at a specific flowmeter setting.
2. Measure inlet and outlet pressures of pump at several flow rates.
3. Record the pressure drop across the flow meter devices at several water flow rates.
4. Record the pressure drop across the various pipe fittings and valves at different water flow rates.

Results:

Calibration curves were determined for the flow meters used in system A. Characteristic constants were also determined for the flowmeters used. Pump characteristic curves were plotted for the pump. The maximum efficiency for the pump was found to be 30%. The results were comparable with published results for similar systems.

The results found for the valves and fittings in the second piping system were the equivalent length and velocity head lost. The equivalent length of a fitting is the length of straight pipe that would result in the same pressure drop as what was measured across the fitting or valve. Velocity head lost is the kinetic energy that is lost by the fluid flowing through the fitting. The results found allow for accurate prediction of the effects on a fluid flowing in a piping system.

(handwritten left margin) A LITTLE TOO GENERALIZED- NEEDS A BIT MORE DETAILS TO AVOID CONFUSION

E.Q. THREE DIFFERENT FLOW METERS

Conclusions

1. The flowmeters used are accurate for measuring flow rate in pipes.
2. Pump efficiency increases with increasing flow rate.
3. The pump used was oversized for the both piping systems.
4. Equivalent lengths and velocity head lost are accurate methods for predicting friction losses in valves and fittings.

(handwritten) IN GENERAL, O.K. BUT A LITTLE TO VAGUE

(handwritten) MAKE SUBHEADINGS STAND OUT MORE

Assessment Criteria for the Technical Translation Memo

Below are listed the criteria for a *technical translation memo*. After reading the exhibit, please circle the response indicating your judgment as to whether the group has exhibited.

4 - Superior ability

3 - Competent ability

2 - Somewhat competent ability

1 - Lack of competent ability

Lead

1. The aim and significance of the research have been identified.

4   3   2   1

Apparatus

2. A block diagram of the apparatus has been provided. This schematic illustrates only the core principles that underlie the experiment.

4   3   2   1

3. The procedure by which the equipment was operated has been provided, but only the most significant steps have been provided.

4   3   2   1

Results

4. The experimental findings have been clearly presented to indicate relationships of variables.

4   3   2   1

5. There is an explanation of why those findings are of importance.

4   3   2   1

Conclusion

6. An interpretation of the research is provided in terms of the most important aspects of the experiment.

4   3   2   1

Discussion Questions on the Technical Translation Memo

1. Consider the translation strategies the writers employed in the exhibit. Were they successful in getting across the essence of the work performed? If not, what could have been done?

2. In the exhibit, the integration of a basic concept (flow=driving force/resistance) is very effective. Why?

3. The basic concept mentioned in Question 2 has application in numerous disciplines as well as various areas in chemical engineering. Discuss as many of these specific cases as possible and their relative usefulness as analogies.

4. Why is the schematic especially useful? Is it detailed enough for a technical translation? If not, what else is needed?

5. The Results section is so general as to be ineffective. What level of additional detail is needed?

6. Go to your local library and review the best-seller list for non-fiction for the last month or so. Select one book on science and read it through, noting the kinds of techniques the author used in translating the information in the book for a popular audience.

7. Have you had experiences with presenting information to various audiences in your other courses. If you have, how were these assignments structured? Were they useful? If you have had no such experiences, what do you think might be the reason that your curriculum does not address technical translation?

---

[1] A. Einstein, "E=MC²," *Science Illustrated,* 1946. Rpt. in R. E. Lynch and T. B. Swanzey, *The Example of Science: An Anthology for College Composition,* New Jersey: Prentice-Hall, 1981.

[2] For more on translation see the following: Susan Bassnett-McGuire, *Translation Studies,* London: Methuen, 1980; J. C. Redish, R. M. Battison, and E. S. Gold, "Making Information Accessible to Readers," in L. Odell and D. Goswami, eds., *Writing in Nonacademic Settings,* New York: Guidford, 1985, pp. 129-153. For the best explanation of translation strategies, one that has influenced the above discussion, see David. A. McMurrey, *Processes in Technical Writing,* New York: Macmillian, 1988, pp. 204-235.

The Scale-Up Memo

Introduction

"'Scaling up,'" write John A. Heitmann and David J. Rhees, "is a phrase that sums up the work of the industrial chemist and chemical engineer. It is an activity that transforms the chemist's knowledge of reactions taking place in test tubes, beakers, and small vessels into efficiently designed processes, utilizing pumps, pipes, vats, filters, autoclaves, and other large-scale apparatus." [1]

The authors also quote a former director of research for General Motors who labeled the scale-up process as "the shirt-losing gap." Without scale-up, there would be no chemical engineering. The search for synthetic processes, petroleum cracking, and penicillin production are all excellent examples of the complexities and risks surrounding the scale-up process.

Because it simulates the kind of high-stakes decision making that is part of chemical engineering, the scale-up memo has become to us one of the most challenging reporting structures. Even under the simulation limits of a classroom setting, you will find that it is nerve-shattering to scrutinize data under the pressure of a scale-up decision made in an industrial setting. How good, we frequently ask our students, are your results in economic terms? Are you willing to put into the organization a plan that may, in time, cost $1 million? Are you willing to risk your place in the organization? To risk your place in your profession? Before the scale-up exercise, students perhaps think of *Shreve's Chemical Process Industries* as just another chemical engineering textbook; after the exercise, students begin to realize the complexities of an industrial decision to proceed from a test tube to a production plant. [2]

Organization of the Scale-Up Memo

Of course, the scale-up process is a complex one, involving teams of corporate decision makers. We cannot hope to simulate that environment in this humble exercise. Nevertheless, the stimulus behind the scale-up decision, as Heitmann and Rhees's historical sketch reveals, often derives from either an individual or a small group of researchers. While the implementation of the scale-up process may involve a bureaucracy, the initial research involves individuals like you. So, it is possible to simulate this kind of initial core decision making with this exercise.

The scale-up memo can accompany the traditional laboratory report, the request for funding, the scholarly paper, or the oral presentation proposal. An outline of an effective organizational pattern appears below.

Organization of the Scale-Up Memo

Introduction

In the first sentence, state directly whether or not you recommend scale-up. Establish categorically the reasons for your recommendations.

(Tactical note: Either you will recommend that scaling-up be undertaken or that it be delayed because of uncertainties. In either case, identify your reasons categorically from the least important to the most important. You will explicate these reasons in the *Discussion* section.)

Background

Identify the problem at hand.

Identify what you have been asked to do.

(Tactical note: In this section, provide the context for the demand for scale-up. Think in terms of the questions posed for historical analysis in Chapter 2.)

Procedure

Discuss the strategy by which you have attempted to solve the problem under discussion by briefly describing your apparatus and procedure.

Provide a figure for the apparatus.

Discussion

Identify the reason for your claim in categorical form.

(Tactical note: Remember that you are presenting the heart of your argument here, so use the strategy suggested by Toulmin's scheme in Chapter 4. Remember also that the reader must be able to isolate the supporting reasons for your claim because this is the section in which you explicate the categorical reasons you identified briefly in the introduction.)

Recommended Actions

Specify the next steps to be taken in a procedural format.

(Tactical note: Regardless of your decision, in this section you must give a direction that should be followed. Never, never lead a reader into a problem and fail to provide a way out of that problem.)

Description of the Experiment

The exhibit is based on an experiment in which liquid flow through piping and fittings is studied. The primary objectives include: 1) the determination of the number of equivalent lengths and equivalent velocity heads lost by flow through various fittings, 2) the determination of the friction factor as a function of Reynolds number for straight pipe flow, 3) evaluation of orifice and venturi meter flow coefficients, and 4) evaluation of the characteristic curve for a centrifugal pump.

252

To:
From:
Date :
Re:

Introduction

Scaling up the fluid flow experiment is not presently
recommended. More experiments need to be performed in order to
make a qualified decision on scale up.

1. Due to the lack of technical data, it is unknown if the
required equipment can be made to the desired larger scale.
2. Scale up of this experiment may not be an economically sound
proposition due to increased capital and operating costs.
3. The most fundamental reason against scale up is the fact that
only one set of experimental parameters were tested. Sensitivity
analyses on different parameters were not investigated.

FAIRLY STRONG QUALIFICATION WHY SO PESSIMISTIC?

Background *e.g.*

Scale up includes the enlargement of the pipe diameter, fittings,
pump and metering devices. The characteristics of fluid flow
thru these devices were examined in this experiment. The
importance of this examination is to establish a basis for piping
systems in industrial plants, i.e. waste water treatment
facilities. In a waste treatment facility, large quantities of
fluid must be transported efficiently to the necessary areas.
The characteristics of pipes and fittings utilized in the sewage
plant must be large enough to transport the quantity desired,
durable enough to withstand long term usage, and to exhibit
minimal pressure head loss or resistance. Minimization of head
loss is significant in reducing the size and power of the pump,
thus minimize capitol and operating costs.
ing

Procedure

The experiment studied fluid flow thru a 1/4" stainless steel
pipe which was fitted with orifice and venturi meters, gate,
globe, lift check and swing check valves, angle and straight
tees, 45° and 90° elbows, and a 10 ft. run. Figure 1 illustrates
the system investigated. The head losses due to the fluid flow
across each of these resistances were measured by a manometer.
This was done to gain knowledge of possible power requirements.

Discussion

AS MUCH AS

The pipe diameter investigated in this experiment was only 1/4",
but a scaled up model might be five feet in diameter. Larger
pipes are constructed from materials other than stainless steel
to endure larger mass flowrates and hostile environments. Sewer
lines tend to be fabricated from cement in order to lower
maintenance costs and to withstand extreme operating conditions.

RESIST CORROSION   NO!

NOT NECESSARILY

253

The type of pump may be varied to gain efficiency in scale up. The centrifugal pump used in this case will not have the capacity for enlargement.

A straightforward scale up of our experiment would not be profitable for large industrial plants. The manufacturing and operating costs would be non-realistic. A scaled up system would transport greater mass flow over longer distances. This increase requires alterations of the following parameters: pipe length and diameter, fluid velocity and pump requirements.

Scale up of these parameter would not be in equal proportions to the mass flowrate. An increase in the diameter by a factor of 2 would result in a velocity decrease of 25 %. Another scale up problem involves fluid density and viscosity. In our experiment, the fluid used was water. In a sewage plant the effluent has extremely different properties. This change is an important consideration due to its affect on the Reynold's number.

Recommendations

The following recommendations would improve analysis for a scale up design:

1. Multiple runs of our experiment and all future experiments should be performed for precision.
2. Different diameter systems should be experimentally analyzed.
3. The experiment should be repeated for pipes constructed of various materials, i.e. glass, cement, plastic, etc..
4. The fluid type should be altered for a larger range of density and viscosity values.
5. After analyses of the above experiments, a pilot plant scale up model should be developed and tested.
6. The results of the pilot plant runs would either confirm or reject scale up feasibility.

NOT BAD! Good in light of introduction — MAYBE SCALE-UP WILL BE POSSIBLE LATER

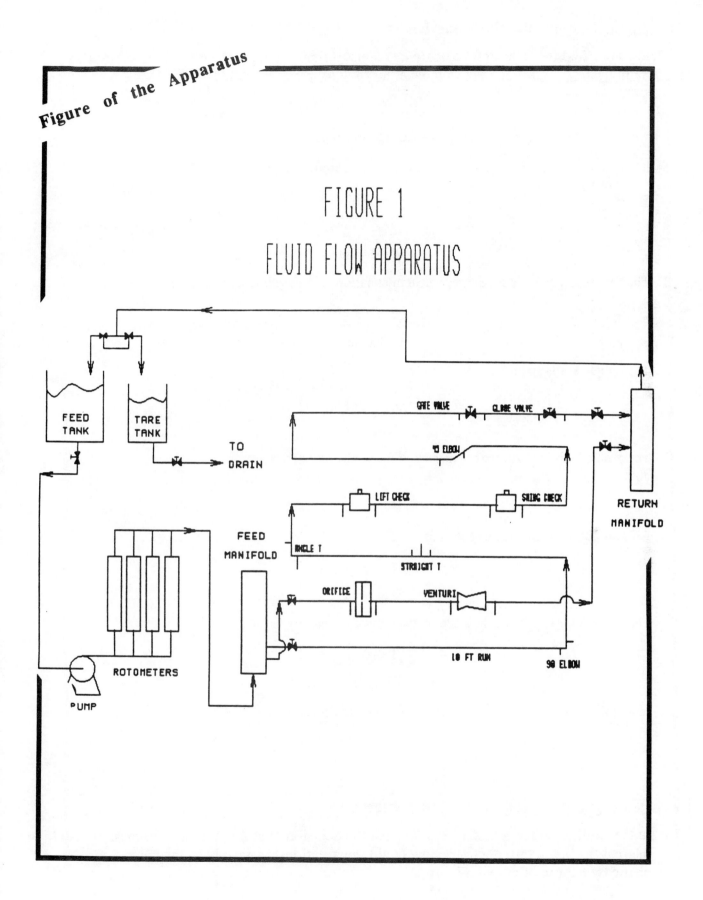

FIGURE 1

FLUID FLOW APPARATUS

Assessment Criteria for the Scale-Up Memo

Below are listed the criteria for a *scale-up memo*. After reading the exhibit, please circle the response indicating your judgment as to whether the group has exhibited

4 - Superior ability

3 - Competent ability

2 - Somewhat competent ability

1 - Lack of competent ability

Introduction

1. The first sentence includes a direct recommendation for or against scale-up.

4   3   2   1

Background

2. The unique scale-up problem is identified.

4   3   2   1

Procedure

3. The problem-solving strategy is discussed by means of a description of the apparatus and procedure.

4   3   2   1

4. A figure for the apparatus has been provided.

4   3   2   1

Discussion

5. The reasons for or against the scale-up decision are cogently argued.

4   3   2   1

Recommended Actions

6. In a procedural format, the next steps to be taken are delineated.

4   3   2   1

Discussion Questions for the Scale-Up Memo

1. In the exhibit, do the reasons for or against the decision to scale up in the Discussion match the reasons in the Introduction? How could this relationship have been made more emphatic by the use of graphics?

2. Using the Toulmin scheme presented in Chapter 4, diagram the argument in the exhibit. Are there more supporting pieces of evidence of more qualifications? What does this relationship tell you about the balance of proof in a scale-up decision?

3. Many of the most recent scale-up decisions have been made by the use of computer-aided design. Locate an instructor in your department who uses computer modeling to design components such as heat exchangers and reactors in laboratory planning. Interview the instructor about the type of computer software used, the strengths and limits of that software, the strengths and weaknesses of computer-aided design, and the industries that most use this new technology. Make a ten-minute oral presentation to your class about your findings from this interview.

---

[1] J. A. Heitmann, D. J. Rhees, *Scaling Up: Science, Engineering, and the American Chemical Industry*, Philadelphia: Center for the History of Chemistry, 1984.

[2] G. T. Austin, *Shreve's Chemical Process Industries*, 5th ed., New York: McGraw-Hill, 1984.

The Formal Laboratory Report

Introduction

You are probably very familiar with the traditional laboratory report. It incorporates the very process of engineering research discussed in Chapter 3. When did this rhetorical form—so common in the laboratory curriculum—emerge?

If you look for the answer in the standard history textbooks, you would be led to believe that the scientific article—upon which the laboratory report is based—emerged with the scientific method during the 17th century. Yet a recent study by Charles Bazerman finds that the scientific article as we know it is a recent phenomenon.[1] After examining the *Philosophic Transactions of the Royal Society of London*, Bazerman identified four stages in the history of the experimental report. In the first stage (c. 1665–1700), articles simply consisted of uncontested reports of events. In the second stage (c. 1700–1760), experimental articles took the form of arguments over experimental results. In the third stage (c. 1760–1780) articles concentrated on the process of discovery. Only in the fourth stage (c. 1790–1800) did the experimental article reach the form in which we know it, complete with claims and experimental proofs.

What do Bazerman's findings indicate about the ways that research is reported?

First, even though we would like to believe that the experimental article on which the laboratory report is modeled has been around since Galileo, the evidence points to the fact that this reporting structure is really quite recent, only a little more than 200 years old.

Second, once we realize that the form of reporting structures varies across time, we realize that the laboratory report is only one of many ways that experimental research has been reported. The current form is quite recent. Indeed, given the organizational settings in which you will work after graduation, the laboratory report may be a document that you will rarely encounter outside of the university.

Third, we may then infer that, although the kind of engineering research process discussed in Chapter 3 does not vary, the method by which it is reported does. An organizational reporting environment is far different from an academic one. It therefore seems prudent for you to explore as many reporting structures as you can so that you may be better prepared to adapt your reporting structures to the contexts in which your research will be conducted.

None of this is to say that the laboratory report is not a valid way to present information. An academic genre, it can provide you with a way to explore your ideas much more freely than the restricted briefing memo or scale-up memo. It can also inspire you to pursue the academic career path of engineering in which the scholarly paper becomes important.

The key to learning to communicate effectively, as we have stressed throughout this book, is for you to experiment with multiple perspectives of thought. This may be accomplished while using the academic laboratory report as the cornerstone of your laboratory course.

Organization of the Laboratory Report

The laboratory report can be the required structure for all work in the laboratory, or it can be one of many reporting structures. An outline of the organizational pattern that is traditionally followed is provided below.

Organization of the Laboratory Report

The Title Page

List the title, institutional affiliation, group number, members of the group, and the date.

Table of Contents

List all of the sections in the report with the correct page designation.

Make sure that the order of the sections follows the order in the report.

Make sure that the exact wording is used for the table of contents that is used in each section of the report.

Abstract

Summarize the major points of the report in one paragraph.

(Tactical note: [1.] In their *Handbook of Technical Writing*, C .T. Brusaw, G. A. Alred and W. E. Oliu identify what they call descriptive abstracts. [2] This kind of abstract includes information about the purpose, scope, and methods used to arrive at the findings of the original document. As such, it is quite like an expanded table of contents in sentence form. Use this type of abstract in the laboratory report. [2.] Both students and colleagues tell us that this is often the most difficult part of the report, so great care must be taken to ensure that it includes only the highest level of information presented in an order that parallels the organization of the report. [3.] Write the abstract after you have completed your draft of the report. You cannot possibly write the abstract until you have let your writing work its power over your ideas. Only when you have the experiment firmly reified by means of the laboratory report should you attempt to write the abstract. Even then you should have it peer reviewed before submitting it to your instructor.)

Introduction/Theory

Relate the experimental problem to the engineering problem under consideration, preferably using only a paragraph-length statement.

Analyze the theoretical aspects of the problem.

Integrate these theoretical aspects to reveal a qualitative-quantitative relationship.

Include a schematic to display the research problem visually.

Objective

Establish the aim of the experiment in a complete, lucid sentence.

(Tactical note: This statement should be drafted early in the planning stage of the experimental process. Along with the abstract, this may be the single most difficult section to write.)

Procedure

Describe the process used in obtaining the experimental data.

Provide a neat sketch of the apparatus.

(Tactical note: Recall our discussion in the guide for the laboratory notebook that anyone should be able to repeat your experiment and make similar observations. Be sure that this amount of detail is provided in this section as you describe your experimental procedure.)

Description of the Apparatus

Provide a discussion of any special features of the equipment.

Provide key dimensions and operational rates that influence the results.

Results

Present your results as clearly and objectively as you can.

(Tactical note: Present your results separately from your discussion. In this way, you will be able to separate your experimental findings from your interpretation of those findings.)

Discussion

Explain and justify your analysis of the results you have just presented.

Discuss numerical results, graphical presentations, comparisons with theoretical expectations, and equipment limitations.

Discuss important results of the experiment in reference to key tables and graphs.

(Tactical note on tables and graphs: Because you want your reader to interpret these visual representations of your research in the same way that you have, follow this three-step method. Step 1. Introduce the figure into the text by identifying it with a number; Step 2. Present the table or graph clearly and accurately, remembering to give it an accurate title; Step 3. Explain what the reader is supposed to infer from the table or graph by taking your reader through the relationships expressed in the data.)

(Tactical note: Remember that this section is the heart of your engineering argument, so you might make a sketch of your argument according to the Toulmin model [see Chapter 4] in your laboratory notebook before you attempt to write this section.)

Conclusions

Provide a coherent picture of the work accomplished by commenting on qualitative/quantitative relationships.

(Tactical note: To ensure cohesiveness in the report, make this section as parallel as possible to the Objective section. For clarity, it is often good to set your conclusions off numerically.)

Suggested Modifications to the Lab

Itemize needed equipment repairs.

References

Provide a list of all pertinent published reference sources.

(Tactical note: Check with your department or graduate school for the required format of references. Be sure you cite your references accurately so that others may follow your research. Hence, you should take as much care with this section as you do with the Procedure section.)

Appendix

Provide sample calculations.

Include the appropriate levels of data discussed in Chapter 4 of the following: experimental data, tables, graphs, and computer data.

Provide a brief table of contents to allow readers to locate material in the appendix easily.

Description of the Experiment

In this experiment, the performance of a packed gas-liquid absorption tower is evaluated. The primary objective is to determine how the mass transfer rate is affected by gas flow rate, especially as the column approaches its loading and flooding points.

The absorption of NH_3 from an air stream into a counter-current water flow is an effective system for the study of mass transfer in a packed tower. The important concepts of overall and individual film mass transfer coefficients, transfer units, closure, loading, and flooding are all considered.

Abstract

PERFORMANCE CHARACTERISTICS OF PACKED TOWERS

Department of Chemical Engineering
New Jersey Institute of Technology

Laboratory Group #1

Submitted:

ABSTRACT

The performance characteristics of Spherical and Intalox packings, such as column pressure drop, were determined for both dry and wet operations. With a single fluid (air) passing through dry packing, the pressure drop was found to be directly proportional to the fluid mass flow rate. This observation was found to be independent of the packing used.

With two fluids (air and water) flowing countercurrently to one another, it was determined that Intalox saddles were a superior packing than spheres. Unlike spheres, Intalox saddles had a high surface area per unit volume, thus providing good liquid distribution and a larger contact area between the two fluids. Furthermore, it was proven that under similar operating conditions, columns packed with Intalox saddles rather than spheres, experienced smaller pressure drops and realized higher flooding velocities.

Finally, under similar operating conditions, the Intalox saddles had a larger liquid hold-up due to their larger porosity value and smaller size.

Too Speculative For Abstract

Generally Good — To The Point

262

1

INTRODUCTION

Packed towers are typically used for absorption in the continuous countercurrent contacting of a gas and a liquid. Gas enters below the packing in the column and rises upward through the openings or interstices of the packing and contacts the decending liquid flowing through the same openings. A large contact area between the gas and the liquid is provided by the packing. Many different types of packing have been developed, each providing a different void space. High void volumes of 60 to 90% are characteristic of good packings.[1]

Good use of reference →

The purpose of this experiment was to identify which packing, either the spherical or Intalox saddles, provided superior performance through determination of the hydraulic characteristics for each type of packing.

WHAT ARE THEY?

THEORY

Consider a column filled with a dry packing of an effective diameter, Dp. A geometric constant, known as the porosity, can be defined as[2]:

$$\epsilon = \frac{\text{volume of voids in column}}{\text{total volume of column}} \qquad (1)$$
$$\text{(voids and packing)}$$

This constant, ϵ, and the effective diameter, Dp, can be related to the hydraulic radius by[3]:

$$r_H = \epsilon/[6(1-\epsilon)] \, Dp \qquad (2)$$

which in turn, is related to the equivalent diameter by:

$$r_H = 1/4 \, (D) \qquad (2a)$$

The value of ϵ is characteristic of the particular packing used, and depends upon the shape and size distribution of the packing, the ratio of packing size to column diameter, and the method used in distributing the packing inside the column (ie. random or stacked).

If a fluid, such as air, is flowing at an average velocity V through the column, the superficial velocity V' may be defined as[4]:

$$V' = \epsilon V \qquad (3)$$

263

2

The superficial velocity is the velocity in the open section below the packing support.

Knowing the physical properties of the fluid, an expression for the Reynolds number can be derived as[5]:

$$N_{Re} = \frac{DpV'p}{(1-e)\mu} \qquad (4)$$

For laminar flow ($N_{Re}<10$), the Hagen-Poiseuille equation can be combined with equations 3 and 2 to give[6]:

$$\Delta P = \frac{72\mu V'\Delta L(1-e)^2}{e^3 Dp^2} \qquad (5)$$

Where ΔL is the height of the packing in the column[7].

Experimentally, it has been show that the constant should be 150, thereby giving[8]:

$$\Delta P = \frac{150\mu V'\Delta L}{Dp^2} \frac{(1-e)^2}{e^3} \qquad (6)$$

This indicates that for a given system, the flow rate is directly proportional to the pressure drop.

For turbulent flow ($N_{Re}>1000$), the Hagen-Pouiselle equation can be combined with equations (2) and (3) to give[9]:

$$\Delta P = \frac{3fp(V')^2\Delta L}{Dp} \frac{1-e}{e^3} \qquad (7)$$

It is assumed that for highly turbulent flows, the friction factor, f, approaches a constant value and it is also assumed that all packed columns should have the same relative roughness. Experimentally, it has been shown that $3f=1.75$, thereby giving[10]:

$$\Delta P = \frac{1.75p(V')^2\Delta L}{Dp} \frac{1-e}{e^3} \qquad (8)$$

For flows in the transitional region $1<N_{Re}<1000$, equations (6) and (8) can be combined to give the Ergun equation expressed as[11]:

$$\frac{\Delta Pp}{G'^2} \frac{Dp}{\Delta L} \frac{e^3}{1-e} = \frac{150(1-e)}{N_{Re}} + 1.75 \qquad (9)$$

where $G'= V'p$.

3

The first term on the right in equation 9 represents the fractional viscous energy loss; the second, the fractional kinetic-energy loss. Ordinarily the former is relatively small, the latter large[12].

Consider the case in which liquid and vapor phases are involved. The liquid descends through the column in the form of films distributed over the packing surface, and the vapors rise through the spaces between the packing particles.

Under operating conditions, the column has the capacity to hold a quantity of liquid between the spaces of the packing and on the packing surface, which is appropriately termed the liquid hold-up. The quantity of liquid usually increases with increasing gas velocities and may be estimated by[13]:

$$h = K(L_1)/Dp^n \qquad (10)$$

As liquid holdup increases rapidly with the gas flow rate, the free area for gas flow becomes smaller and the pressure drop rises more rapidly. This is known as loading[14]. For packed beds under pre-loading conditions with simultaneous counterflow of liquid and gas, the empirical equation by Leva may be used to estimate pressure drop[15]:

$$\frac{\Delta P}{L} = \alpha (10)^{\beta L} \frac{G^2}{p_v} \qquad (11)$$

where L and G are liquid and gas mass fluxes, p_v is the vapor density, and α and β are experimentally derived constants.

If the gas velocity is increased beyond the loading condition, the column will begin to fill with liquid, and the gas will bubble through the spaces between the packing particles. The column will appear to be filled with a "boiling" liquid, and the transition from a gas continuous liquid dispersed situation to a liquid continuous gas dispersed situation has occurred. This condition is known as flooding, and represents the upper limiting conditions of pressure drop and fluid rates for practical tower operation. Flooding velocities can be correlated by the relationship[16]:

$$(G_v^2 A_s \mu_L \cdot^2)/(g_c G^3 p_v p_L) = f\ [(L_1/G_v)(p_v/p_L)^{1/2}] \qquad (12)$$

Again, the gas flow rate calculated by the above equation is indicative of the column's maximum limitations.

GOOD THEORY SECTION
- KEY EQUATIONS ALL INCLUDED
BUT, I PREFER ALL NOMENCLATURE LISTED IN TEXT, AS FIRST USED.

4

OBJECTIVE

The purpose of this lab was to determine the hydraulic performance characteristics of Spherical and Intalox packings for both dry and wet operations.

For the dry operation, the influences of viscous and kinetic forces were evaluated using the Ergun equation. Furthermore, the pressure drop across the column was determined as a function of air mass flow rate.

The dynamic properties of wet packing, such as the variation of liquid hold-up with gas rate, the determination of loading and flooding points, and column pressure-drop, were also investigated. Theoretical predictions of liquid hold-up and pressure-drop were then compared to experimental values.

Good
Thorough

PROCEDURE

Two columns were investigated; one containing spherical packing, the other Intalox saddles. Operation of the packed absorption towers involved measuring pressure drops across the tower at dry and wet conditions. Figure 1 shows a schematic of the packed towers used in this experiment.

Figure 1
Packed Absorption Tower

Good Schematic !
Detailed Enough

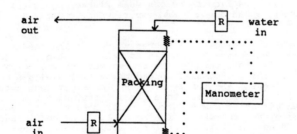

R : Rotameter

▓ : Pressure tap

5

Initially, the operating characteristics of dry packing were examined. A known air flow rate at a relatively low ← TOO WORDY
rate was allowed to flow into the bottom of the column containing the spherical packing using a rotameter as a metering device. The pressure drop across the column was then measured using a manometer. This procedure was repeated using the full range of the rotameter (1.945 acfm-30.534 acfm), measuring pressure drops across the column as functions of air flow rate. The column containing Intalox saddles was examined using the same technique.

The hydraulic operating characteristics of wet packing were studied next. A relatively low water flow rate (.395 gpm) was allowed to flow into the top of the column. Using the full range of rotameter, air was allowed to flow countercurrently with the water, and pressure drops were measured across the column as functions of air rate.

For each gas rate, the liquid hold-up was measured. During steady state operation, the liquid level at the bottom of the column was noted. By stopping the liquid flow into and out of the column, liquid from the packing was allowed to drain until the liquid level at the column base stopped rising. The liquid from the column was then drained into a tared vessel until the liquid level reached the original steady state position.

The pressure drops and liquid hold-ups were then obtained using the previously mentioned procedures for relatively high water rates (.624 and .78 gpm). The column containing the Intalox saddles was analyzed in an analogous manner. After completion of all experiments, the main water and air supply lines were shut down and all systems drained.

6

RESULTS

Dry Packing

For both columns containing dry packing, the pressure drop per foot packing, $\Delta P/L$, was plotted against the vapor mass flow rate per unit of column cross section, Gv. This resulted in two linear relationships between $\Delta P/L$ and Gv for both spherical and Intalox packings (See Figures 2 and 3).

Next, the Ergun equation was evaluated. Substituting the appropriate values into equation (9), and plotting the left hand side of the Ergun equation versus $(1-\epsilon)/N_{Re}$ yielded linear relationships (See Figures 4 and 5). The straight lines were of slopes of 626.2 and 1608.1 and their corresponding intercepts were 3.36 and 32.2 for the spheres and saddles, respectively. These values were in disagreement with equation (9), which predicted a straight line with a slope of 150.

Wet Packing

For the column containing spheres, at water flow rates of 0.39 and 0.624 gpm, the logarithm of $\Delta P/L$ was plotted against the logarithm of Gv (See Figures 6 and 7). These plots differed from those of the dry packing in that there were noticeable changes in the slopes of the lines. These breaks were indicative of loading and flooding points. For water flow rates of 0.39 and 0.78, the same type of plots were constructed for the column containing Intalox saddles, and similar trends in slope deviation were noted (See Figures 8 and 9).

The liquid hold-ups for each respective packing were correlated using equation (10). For two liquid mass flow rates per unit of column cross section, the following relationships may be derived:

$$n = [\log(h_1/h_2)]/[\log(Ll_1/Ll_2)] \qquad (11)$$

and

$$\log K = \log h - \log (Ll/Dp)^N \qquad (12)$$

Where h_1 and h_2 are the liquid hold-ups at liquid rates Ll_1 and Ll_2, respectively. Solving the system of equations (11) and (12) yielded values for n and K, for the spherical and Intalox packings (See Table 1).

268

7

Table 1
Liquid Hold-up Correlations

Packing	n_{exp}	k_{exp}	n_{lit}	k_{lit}
Spheres	.865	.00032	.60	.00040
Intalox Saddles	.604	.00024	.55	.00040

exp: experimental values
lit: literature values[13]

The packing factors α and β for each respective packing were determined using equation (11). Again, for two liquid mass flow rates per unit of column cross section, Ll_1 and Ll_2, the following relationships may be derived:

$$\beta = [\Delta P_2/\Delta P_1][Ll_2-Ll_1] \qquad (13)$$

and

$$\alpha = (\Delta P/L)/[10^{BLl}(Gv^2/p_v)] \qquad (14)$$

where for a given air mass flow rate Gv, ΔP_2 and ΔP_1 are the pressure drops at liquid mass flow rates Ll_2 and Ll_1. Solving the system of equations (13) and (14), yielded values of α and β for the spherical and Intalox packings (See Table 2).

Table 2
Pressure Drop Correlations

Intalox		Spheres	
Experimental	Literature[17]	Experimental	Literature
α = .25	α = .82	α = 1.0	α = ?
β = 5.75	β = .20	β = 30	β = ?

* WHERE ARE THE UNCERTAINTY LIMITS (BARS) FOR THE EXPERIMENTAL QUANTITIES?

DISCUSSION

Concerning dry operation, the observed linearity in Figures 2 and 3 is an indication of the proportionality between the pressure drop across the bed and the air mass flow rate. This observation is independent of the packing used.

The Ergun correlations, as seen in Figures 4 and 5, show substantial deviation from the theoretically predicted slope of 150. Reasons for this discrepancy are provided by McCabe, Smith and Harriot. Firstly, the Ergun equation predicts pressure drops lower than those found experimentally. Secondly, both columns were probably wet from a previous wet run done by another group of experiments. If so, the water may be considered another resistance for air flow, thus translating into a greater pressure drop. Since the pressure drop term is present in the left side of the Ergun equation, an increase or decrease in slope occurs. Therefore, both of these factors would result in a slope much larger than theoretically expected.

WEAK!

MORE LIKELY TO BE OLD PACKING- MECHANICAL ATTRITION

For wet operation, the pressure drops for spherical and Intalox saddle packings may be represented by the following relations:

$$\Delta P/L = 1.0 \times 10^{30L}(Gv^2/p_v) \qquad \text{Spheres}$$

$$\Delta P/L = .25 \times 10^{5.75L}(Gv_f^2/p_v) \qquad \text{Intalox}$$

Since limited information was available for spherical packing, comparisons between theoretical and experimental pressure drops could not be made. However, considering the Intalox saddles, α and β and thus ΔP did not compare

270

IS THIS VALUE REALLY THAT BAD?

19

favorably with predicted literature values, differing by as much as 88% (See Figures 8 and 9).

Liquid hold-ups for the spheres and saddles were found to be represented by the following relations:

$$h = .00032(Ll/Dp')^{.865} \quad \text{Spheres}$$
and
$$h = .00024(Ll/Dp')^{.60} \quad \text{Saddles}$$

The theoretically predicted hold-ups for spheres compared quite favorably to the experimentally determined hold-ups, differing by an average of 21%.

Unfortunately, the liquid hold-ups for the Intalox saddles did not correlate well, deviating by an average of 78%.

DON'T APOLOGIZE! QUANTIFY

Flooding velocities were determined by plotting values calculated by equation 12 directly on the flooding curve by Leva[15]. It was noted that for increasing water mass flow rates, the flooding velocities decreased. This observation was independent of the packing used. Furthermore, for similar water mass flow rates, a greater velocity of air was needed to flood the column filled with Intalox saddles, than for spheres. This indicates that a column filled with saddles will realize its maximum limitations at much greater air velocities, than a column filled with spheres.

The Effects of Channeling

Channeling may be considered a departure from the uniform distribution of the phases in a packed column, and usually occurs when the packing geometry inhibits lateral distribution. Although there are no methods for predicting the amounts of channeling, this condition appears to be favored by column diameter/packing diameter ratios greater than 30[11]. When the systems of spheres and saddles were considered, the resulting ratios were 12 and 18.75, respectively. Therefore, for this particular experiment, channeling was not a significant problem since the criteria was not met.

GOOD - DIFFERENT!

Relative Contributions of Viscous and Kinetic Energy Losses as Functions of Flow Rate

For the flow of a single fluid phase through a bed of packed solids, the viscous forces control at low N_{Re}. Therefore, equation 6 is applicable. However, at higher N_{Re}, inertial forces control, and thus equation 7 is valid. For intermediate N_{Re} ($1<N_{Re}<1000$), the viscous and inertial forces are additive, thereby resulting in the validity of the Ergun equation (Equation 9).

YOU DON'T REFER TO YOUR DATA HERE

271

20

THIS
TEXT IS
OK FOR
A TEXTBOOK,
BUT NOT
A LAB REPORT.

Physical Occurrences at Flooding and Loading Points

It is usually advantageous to maximize either of the
flow rates in a two phase (liquid/gas) system. If either
the liquid rate or gas flow rate or both are increased, it
is observable at a certain point, that the normally
dispersed liquid can not trickle down through the packing.
Pools or pockets of liquid are seen to form, which reduce
the available area for gas flow. This undesirable
condition, in which the interfacial area for contact is
reduced, is known as loading.

If the gas flow rate is further increased, the liquid
pools formed in loading will consolidate across the entire
cross sectional area of the column. The gas can only bubble
through the liquid, and the column will begin to fill with
liquid, causing the pressure drop across the column to
increase without bound. This unstable condition is known as
flooding, and will continue until the liquid is carried out
over the top of the column with the exit gas.

**Comparison of Spheres and Intalox Saddles on a Basis of
Performance**

Unlike spheres, Intalox saddles were designed to
provide a greater degree of randomness, thus resulting in
lower pressure drops per foot packing. Furthermore, no two
saddles can cover each other to such an extent that large
degrees of contact area are rendered ineffective, due to
mutual screening. This results in excellent liquid
distribution qualities, and superior column stability.

Finally, packing material tended to move from spheres
towards more elaborate designs in order to minimize pattern
packing, reduce column pressure drop and increase gas/liquid
contact area.

Validity of Liquid Hold-up Correlations

For spheres, the theoretical liquid hold-up was predicted by
equation 10 to be 0.05283 ft^3 H20/ft^3 packing. When
compared to actual liquid hold-ups at various gas flow
rates, there was an average deviation of 21.5%. The
correlation for spheres correlated quite well. Considering
the Intalox saddles, the theoretical liquid hold-ups was
predicted by equation 10 to be 0.04874 ft^3 H_2O/ft^3 packing.
When compared to experimentally obtained liquid hold-ups,
there was an average deviation of 78.4%. Obviously, the
liquid hold-ups for Intalox saddles did not correlate well.

Variation of Liquid Hold-up with Gas Velocity

At constant water flow rates, a direct relationship was
noted between liquid hold-up and gas flow rate. In other

21

words, the liquid hold-up tended to increase with increasing
air velocities, provided the water flow was kept constant.
Although no correlation between hold-up and pressure drop
could be found, pressure drop was found to increase with
increasing gas velocities as well.

Validity of Pressure Drop Correlations

For dry packings, the experimental pressure drops did
not correlate well with theoretical predictions. Predicted
slopes of the Ergun equation were in error by 76% and 91%
with the experimental values for spheres and saddles,
respectively. This error was most probably attributed to
the packing being wet and thus providing additional
resistance to flow, which in turn translated into a higher
pressure drop.

AT LEAST, HERE, YOU ARE REFERRING TO YOUR OWN WORK

Due to the limited amount of information available for
spherical packing, pressure drops for wet operation could
not be predicted. For the wet operation of Intalox saddles
the equation presented by Leva (Equation 11) did not
correlate the experimental data well. Predicted and
experimental values were in error by 87.7% (See Figures 8
and 9). The reason for this large discrepancy is the
probable operation in the lower limit of the loading region.

CONCLUSION

· Regardless of packing used, the pressure drop is
proportional to mass flow rate of air for dry operations.

· Compared to spherical packing, Intalox saddles possess
greater randomness and a larger void volume thus translating
into a smaller pressure drop across the packing and larger
air flow rates.

· At similar liquid flow rates, the flooding points for
Intalox saddles are realized at correspondingly higher air
flow rates than for spherical packing.

· At similar conditions, liquid hold-up is greater for
Intalox saddles than for spherical packing due to its larger
porosity value and smaller size.

Assessment Criteria for the Formal Laboratory Report

Below are listed the criteria for a *formal laboratory report*. After reading the exhibit, please circle the response indicating your judgment as to whether the group has exhibited

4 - Superior ability

3 - Competent ability

2 - Somewhat competent ability

1 - Lack of competent ability

The Title Page

1. This page lists the title, institutional affiliation, group number, members of the group, and the date.

4   3   2   1

Table of Contents

2. This page lists all of the sections in the report, followed by the correct page designation.

4   3   2   1

3. The sections in the table of contents appear in the same order as they appear in the report.

4   3   2   1

4. The exact wording used for the table of contents has been used in each section of the report.

4   3   2   1

Abstract

5. The major points of the report have been summarized in one paragraph.

4   3   2   1

Introduction/Theory

6. The experimental problem has been related to the engineering problem under consideration in a paragraph-length statement.

4   3   2   1

7. The theoretical aspects of the experiment have been analyzed.

4   3   2   1

8. These theoretical aspects have been integrated to reveal a qualitative-quantitative relationship.

4   3   2   1

9. A schematic diagram has been provided that displays the research problem visually.

4   3   2   1

Objective

10. The aim of the experiment has been expressed in one concise, lucid sentence.

4   3   2   1

Procedure

11. The basic process used in obtaining the data has been provided.

4   3   2   1

12. A sketch of the experimental apparatus is tied explicitly to the procedure.

4   3   2   1

Description of the Apparatus

13. A discussion has been provided on any special features of the equipment.

4   3   2   1

14. Key dimensions and operational rates influencing the results have been provided.

4   3   2   1

Results

15. The experiment has been presented clearly and objectively.

Discussion

16. The analysis has been explained and justified.

4   3   2   1

17. Numerical results, graphic presentations, comparison with theoretical expectations, and equipment limitations have been provided.

4   3   2   1

18. Important results of the experiment have been provided in reference to key tables and graphs.

4   3   2   1

Conclusions

19. A coherent picture of the experiment is provided by means of comment on the qualitative results.

4   3   2   1

Suggested Laboratory Modifications

20. Any needed equipment modifications have been itemized.

4   3   2   1

References

21. A list of all references used in the report is provided in the required format.

4   3   2   1

Appendix

22. Sample calculations, experimental data, tables, graphs, and computer data have been provided. A brief table of contents is included to help readers locate material in the appendix easily.

4   3   2   1

Discussion Questions for the Formal Laboratory Report

1. Why does the instructor feel that the students' statement regarding Intalox packing is too speculative for the Abstract? Where does a statement like this belong?

2. In a Theory section as complex as that in the Exhibit, why does the instructor prefer that the nomenclature be defined when the terms are first used?

3. Why is the schematic of the pack tower effective? Discuss the figure in terms of which experimental quantities are measured.

4. Contrast the Procedure section of the Laboratory Report with that of the User's Manual and Scholarly Paper reporting structures. What is the relative degree of detail for these three kinds of reports? Explain this contrast in terms of the audience targeted by each structure.

5. Why would uncertainty limits (bars) have ben useful in Tables 1 and 2 of the Exhibit? Might the authors' discussion and conclusions have been different in light of these quantities?

6. In the Discussion, the student authors are still apologizing for their results instead of quantifying uncertainties and any systematic factors which might have affected their results. Comment.

7. The student authors unnecessarily add to the Discussion section. How could the paragraphs noted by the instructor as superfluous have been made more effective?

---

[1] Charles Bazerman, *Shaping Written Knowledge: The Genre and the Activity of the Experimental Article in Science*, Madison: The University of Wisconsin Press, 1988.

[2] C. T. Brusaw, G. J. Alred, W. E. Oliu, *Handbook of Technical Writing*, 3rd ed., New York: St. Martin's Press, 1987, See also C. K. Arnold, "The Writing of Abstracts," *IRE Transactions of Engineering Writing and Speech*, EWS-4, No. 3, 1961.

The Procedural Report

Introduction

In her bibliographic essay on the writing of procedures, Sherry G. Southard reveals that, although technical writing textbooks have included sections on writing instructions since the 1920s, it is only in the 1980s that we find entire books devoted to procedures.[1] Why the sudden and recent growth? At least in part, we may attribute the growth to the rise of the computer industry, an historical phenomenon that necessitated the writing of operator manuals.

Along with the technical translation memo, we see this document as an excellent exercise in audience analysis and variance of genre. To help you design your user's manual with a specific audience in mind and to allow you practice in using the audience analysis technique provided in Chapter 9, we have created the simulation given below:

> You are running the training division in a chemical manufacturing company located in your state. Your corporation employs college seniors each summer to operate a variety of units. These students must operate the units, take measurements, draw conclusions, make inferences, maintain the units, and anticipate problems.
>
> However, your corporation realizes that the quality of undergraduate preparation for such work leaves something to be desired. Therefore, before being sent into the plants around the state, each new group of students is required to operate one experiment in an area such as fluid dynamics, heat and mass transfer, and reactor engineering.
>
> As the training unit managers of this corporation, your task is to design a user's manual for these trainees. It is your job to prepare these trainees before sending them into the field.

Organization of the Procedural Report

Organization of the Procedural Report

Title Page

Provide a title for the manual.

Provide a sub-title that further explains the manual, especially indicating the target audience.

Provide a key illustrative figure that suggests the topic of the manual.

Provide the names of the writers and the date of publication.

Table of Contents

Identify the major sections of the manual and provide page numbers.

Describe briefly what information will be found in each section.

Introduction

Specify how the manual is to be used, and by whom.

Provide an overview of the process under consideration.

Provide an overview of how the results are to be reported.

Background: History, Theory, and Uses of Operation

Discuss briefly the history of the process under investigation.

Discuss the relevant theory of this process.

Identify the uses and significance of this process.

Overview of Training

Specify the major chronological steps to be taken in operating the equipment, reporting the data, discussing the results, and drawing the conclusions.

Procedure I: Operating the Equipment
How should the user prepare for the experiment?
What must be done while the experiment is running?

Procedure II: Reporting Results
How should the results of the procedure be reported?

Procedure III: Discussing Results
How will the user know if the operation is successful?

Procedure IV: Drawing Conclusions
What may be surmised from the experiment?

Trouble Shooting and Maintenance

Anticipate what might go wrong with the experiment and how these problems might be avoided.

Define what maintenance of the equipment is needed.

Index

Specify any important terms used in the manual and where these can be found.

Nomenclature

Specify the important terminology and their definitions.

Appendix

Provide calculations that will be helpful to the user.

As is the case with the scale-up memo, we can only simulate a sense of the difficulties of designing an effective user's manual. There will not be opportunity to design the elaborate field-test procedure that tells writers if the manual truly addresses the needs of the target audience. Nevertheless, the simulation will provide you with an excellent sense of the methods by which you must vary your format to meet the needs of your audience.

The Experiment

The exhibit is based on an experiment in which a York-Scheibel column performing a liquid-liquid extraction is evaluated. Acetic acid is extracted with water from n-hexyl alcohol. The primary objective is to determine how the extraction capability (efficiency) of the column varies with the rate of rotary agitation of its vertical array of stages.

Table of Contents

TABLE OF CONTENTS

No mention of targeted audience

INTRODUCTION

Use of Manual

This manual provides an outline which should be followed for the operation of the Scheibel extraction column. It also serves as a guideline for the necessary analysis to be performed with the experimental data generated.

This procedural manual outlines the steps which must be taken before and during the operation of the unit.

Overview of Process

The Scheibel column is used to achieve liquid-liquid extraction using agitation within the column. This type of extraction is possible due to the chemical potential difference between the phases. This difference provides the driving force needed for mass transfer to occur. By varying the agitation rate and keeping the flow rate of components constant, an investigation of mass transfer versus agitation rate can be performed. The mass transfer should increase due to the increased surface area at high agitation rates.

Good — the user knows what to expect

Overview of Results

The results obtained from this experiment will be reported in the following manner. There will be one graph in the "results" section for each agitation rate studied. These graphs will contain a McCabe-Theile analysis showing the number of theoretical trays needed to complete the observed extraction. From the number of trays, the equivalent height of a theoretical stage will be calculated.

The results will also contain the number of theoretical trays needed for each run using a theoretical equation. If the number of theoretical trays needed decreases the agitation increases, then the experiment has been successful. This data can then be used to predict scale up procedures or to produce a product of a particular mass fraction.

as

Operating the Equipment

It is advised that one should first become familiar with the apparatus. An overall schematic is shown in Figure 3. This column is gravity fed, so the feed tanks must be filled before commencement of experiment. If, at any time, the smell of n-hexanol becomes overwhelming, close the feed tank valves immediately and check for spills.

GOOD SAFETY CONSIDERATION

N-Hexanol and Acetic Acid Mixture to Feed Tank

1. Open valve from storage tank to feed tanks.
2. Turn on pump and visually make sure filling has begun.
3. Fill feed tank to about 3/4 full. Be careful not to overflow feed tank.
4. Shut pump off and shut valve.

Water Feed Tank

1. Open valve from house water supply to feed tank.
2. Turn on house water supply.
3. Fill feed tank to about 3/4 full. Be careful not to overflow feed tank.
4. Close house water supply and shut valve.

Calibration of Rotameters

To establish the flowrates available, the rotameters must be calibrated. Figure 4 demonstrates how to correctly read the rotameter. Note, there are two floats which allow two calibration curves to be constructed.

1. Make sure rotameter valve is fully closed.
2. Open feed valve to rotameters. Also, open purge valve which will allow samples to be taken. Note: Do not open the valve which will allow flow into the column.
3. Open rotameter valve to a marking. Make sure a waste container is placed under the purge valve to avoid spillage. Record marking.
4. Place 1000 ml graduated cylinder under purge valve and note time. Record the amount of fluid collected and time of collection. This will give a flowrate.
5. Steps 3 and 4 are repeated at least 4 more rotameter markings. A calibration curve of markings (verses flow rate can then be generated. This process is repeated for both rotameters.
6. Empty collected waste into waste collection tank.

GOOD DETAIL!

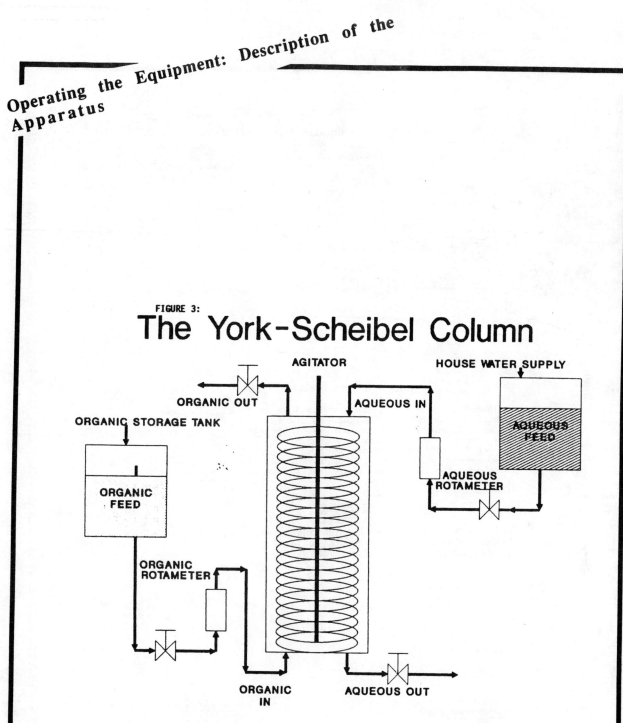

FIGURE 3:
The York-Scheibel Column

GOOD DETAIL!

Safety Precautions

SAFETY PRECAUTIONS

The following precautions should be noted before the experiment is run.

Nice use of block

1) Avoid prolonged exposure to the n-hexanol fumes. Overexposure may cause nausea and dizziness. If any of these symptoms should occur, proceed to a well ventilated area.

2) If a spill occurs, immediately place absorbant upon the spill. Place waste in the proper receptacle.

3) Do not mouth pipette under any circumstances. Use approved suction bulbs provided.

4) Proper safety equipment should be worn at all times while conducting experiments. These include safety glasses and hard hat.

GREAT! Cannot be over-emphasized

Trouble Shooting and Maintenance

TROUBLE SHOOTING & MAINTENANCE

There are several items which must be monitored when operating the column. Failure to check these can lead to erroneous results. It is important to check for leaks in the system. Leaks can be detected visually and by the sense of smell. If this happens, shut the feed valves immediately. A second problem is the feed tanks must be monitored throughout the experiment. The feed tanks must not run dry. If they become low, follow the filling instructions previously given. The third problem involves the fluctuation of flowrates. The can be avoided by periodically checking the rotameters.

There are several things which maybe done to maintain the system. The column must be flushed with water to remove the organics which may cause corrosion. The waste tank must be monitored so that it does not overflow.

USEFUL, ESP. FOR 1ST TIME USER.

Index

INDEX

Assessment Criteria for the Procedural Report

Below are listed the criteria for a *procedural report*. After reading the exhibit, please circle the response indicating your judgment as to whether the group has exhibited

4 - Superior ability

3 - Competent ability

2 - Somewhat competent ability

1 - Lack of Competent Ability

Title Page

1. An accurate title has been provided for the manual, along with a sub-title that further explains the manual.

    4   3   2   1

2. A key illustrative figure has been included that suggests the topic of the manual, and the names of the writers and the date of publication have been listed.

    4   3   2   1

Table of Contents

3. The major sections of the manual have been described briefly.

    4   3   2   1

Introduction

4. Specifications for the use of the manual are provided.

    4   3   2   1

5. An overview of the process under consideration has been provided.

    4   3   2   1

6. An overview has been provided describing how the results are to be reported.

    4   3   2   1

Background: History, Theory, and Uses of Operation

7. The history of the process under investigation has been discussed briefly.

    4   3   2   1

8. The relevant theory of this process has been discussed.

    4   3   2   1

9. The uses and significance of this process have been identified.

    4   3   2   1

10. The major chronological steps to be taken in operating the equipment, reporting the data, discussing the results, and drawing the conclusions have been specified.

4   3   2   1

Trouble Shooting and Maintenance

11. The writers have anticipated what might go wrong with the experiment and how these problems might be avoided.

4   3   2   1

12. Needed maintenance of the equipment has been described.

4   3   2   1

Index

13. Important terms and their locations within the text have been specified.

4   3   2   1

Nomenclature

14. Important definitions have been provided.

4   3   2   1

Appendix

15. Calculations have been provided that will be helpful to the user.

4   3   2   1

Discussion Questions for the Procedural Report

1. Why is it imperative to state "up front" what is the targeted audience for the User's Manual?

2. Why is it so important in a User's Manual for the expected results to be described? Think of user's manuals you have encountered which have been especially good or especially bad. Discuss.

3. Contrast the level of detail in the Procedure section of the User's Manual with those in the Laboratory Report and the Scholarly Paper reporting structures. Discuss the differences in view of the audiences targeted for each.

4. Referring now to the Troubleshooting and Maintenance section of the exhibit, answer question 2 again.

---

[1] S. G. Southard, "Instructions, Procedures, and Style Manuals," in Charles. H. Sides, ed., *Technical and Business Communication: Bibliographic Essays for Teachers and Corporate Trainers*, Urbana and Washington: National Council of Teachers of English and Society for Technical Communication, 1989.

Proposal Request for Funding

Introduction

In her review of current studies of proposals, Alice E. Moorehead finds that, as is the case with procedural manuals, research on proposal writing has increased dramatically over the past twenty years. Once considered technical in nature, proposals are now considered part of a large communications system in which decisions are made from scale-up to marketing.[1]

Implying an argumentative framework, a proposal request for funding compels you to present your research in a manner implying that your ideas have promising consequences that require funding if the profession of chemical engineering is to benefit.

Organization of the Proposal Request for Funding

Organization of the Proposal Request for Funding

Title Page

List the Project Title.

List the investigators' Names and Addresses.

List the Institutional Affiliation.

Table of Contents (one page)

Executive Summary (100 words)

Specify the nature of the research.

Identify the request.

(Tactical Note: The executive summary is an attempt to consolidate the principle parts of a report in one place. Unlike an abstract, the executive summary may be inherently persuasive. Because the executive summary may be read by non-specialists (i.e., budget directors) who may not read your technical report, you should avoid using technical terminology.)

Introduction

Describe the research problem.

Identify the significance of the research and the expected gains.

Provide a precise statement of request.

Background

Provide the historical, theoretical, and industrial relevance of the research.

Identify the literature results.

Provide the results of the experiment at hand.

Proposed Technical Solution

Specify proposed plan of research.

Identify research goals.

Propose expected results.

Management

Describe the fulfillment of research.

Specify a time line for the research.

Specify the kinds of personnel needed to undertake the research.

Requests

Identify requisite new equipment.

Identify requisite materials and supplies.

Identify requisite support services.

Evaluation

Describe how you anticipate your research will be published, presented, review, and evaluated.

References

Provide citations.

Description of the Experiment

The exhibit is based on an experiment in which the hydrolysis of acetic anhydride is studied in a non-adiabatic, non-isothermal batch reactor from which only temperature/time data are taken. The primary objectives include the determination of the Arrhenius parameters (pre-exponential, activation energy) and the heat of reaction from these data only.

Table of Contents

Executive Summary

A method for investigating the kinetics of liquid reactive systems through the use of temperature versus time curves is the focus for present research. The results for this experiment will offer a simple method to calculate the activation energy and pre-exponential factor for the reaction rate constant.

In order to accomplish our goals, it is necessary to update the equipment by adding a computer to receive the temperature signals directly and to automate the calculations. Also, we would like to incorporate a constant temperature bath to the unit. These modernizations will facilitate the experimental procedures greatly.

289

Introduction

Research Problem

In the study of single, homogeneous liquid phase reactive systems, curves of temperature versus time can be used to determine the reaction kinetic parameters. These parameters are the pre-exponential factors and the activation energy. By determining the reaction kinetic parameters by temperature measurements alone, measurements of concentration for the system are not necessary. This point is significant particularly from an economic standpoint, as concentration measurements can be slow and costly.

Once the pre-exponential factors and the activation energy for the reactive system have been established, the rate constant can easily be calculated as a function of temperature. The value of the rate constant is essential to any control application of the reaction. Therefore, running experiments to obtain the reaction parameters is a crucial step in using a reaction in an industrial application.

GOOD MOTIVATION

YOU RUN IT AS PSEUDO 1ST ORDER; RXN IS REALLY SECOND ORDER!

Our particular study is of the hydrolysis of acetic anhydride. The reaction is an exothermic, pseudo first order that runs to completion in about a half hour. Complete cooling of the system after the reaction then takes only about 4-6 hours. The reaction is of added significance because its mechanism is not very well understood. Hundreds of mechanisms have been proposed for it, but none have been confirmed. (Borchardt, H.J. and F. Daniels, 1957, Bottomley, H.J., 1988, Butler, A.R. and V. Gold, 1961.) Often, for this type of reaction, the mechanism is not known, and is thought to be extremely involved and complex. By thoroughly studying the hydrolysis of acetic anhydride, we can better understand it and other reactions like it.

NOT APPROPRIATE HERE

DOUBTFUL

Significance of research and expected gains

VERY GOOD INSIGHT

The primary significance of the research is not the specific study of the hydrolysis reaction but the methodology of the study. As mentioned previously, the exothermic properties of the reaction allows us to determine the kinetic parameters from a curve of reaction temperature versus time. In general, reactions that are best suited to this type of measurement are single homogeneous liquid phase reactions with half-times on the order of 1-5 minutes. Temperature measurement of the reaction is much easier than concentration measurement, requiring only a thermocouple probe connected to a printer instead of the more complex and expensive equipment needed for concentration detection. Therefore, the perfection of the temperature measurement technique will ultimately result in a faster and more economic way to determine the activation energy, pre-exponential factors and the rate constant of these reactions in general.

2

Once the reaction kinetic parameters for a particular system are known, they can be used in the design of control systems for the reaction. When control of the system has been established, the reaction can safely be applied industrially. In this way, our research not only extends the general pool of knowledge on reactive systems similar to the hydrolysis of acetic anhydride, but also saves money and time in industry.

Background

Historical

The ability to determine reaction rates for any system is important to industry. There are many methods for accomplishing this and in general these methods involve the concentration. Finding the concentrations as a function of time for a liquid systems are usually cumbersome and costly. In light of this, this study used a indirect method for the measurement of the kinetic parameters by using temperature versus time values. This experimental procedure is useful for fast, single, homogeneous liquid-phase chemical reactions. Experimental studies similar to the one which was performed have been analyzed by others. (D. Glasser and ID. F. Williams in 1971, R. ID. Williams in 1974 and H. Matusuda and S. Goto in 1984.) Glasser and Williams used a spherical reactor immersed in a constant temperature water bath to study the hydrolysis of acetic anhydride. From their thermal analysis, they were able to determine the heat of reaction and the pre-exponential factor.

GOOD — YOU MUST CONVINCE THOSE WHO WOULD PAY THAT YOUR WORK IS NEEDED BY THEM

Industrial Relevance

This experimental procedure allows one to determine indirectly the measurement for the reaction rate. In industry it is very important to know the reaction rate for a number of reasons. The reaction rate is necessary to determine the size of a reactor vessel. The rate expression is needed so that someone could determine how quickly a reaction takes place so that the effects of a competing reactions can be determined to see weather or not a feasible reaction system can be made. The rate expression is required in the development of control systems.

The method proposed in this experiment offers a procedure by which the reaction rate can be found. The experimental procedure and apparatus is relatively simple compared to some of the other methods. Some of the other methods used to determine reaction rates require some method for sample quenching or the ability to take measurements much faster than the rate of reaction. The proposed method will allow for measurements of temperature to be taken without any interference in the reaction. Also, since only temperature measurements are necessary there is no need for sophisticated analysis equipment to measure the extent of the reaction.

This experimental procedure offers a simple and cost effective manner to determine the activation energy and the pre-exponential factor for the reaction rate constant.

Management

Fulfillment of Research

We will consider our research to be complete when the following standards have been met:

- o All requested modifications to the reactor have been accomplished and are functioning.

- o Enough solid data has been collected to validate the method of temperature measurement analysis. This will consist of data from reactions that fit our criteria, such as the hydrolysis of acetic anhydride and the reaction of hydrogen peroxide and sodium thiosulfate. The data must compare reasonably to established literature values.

GOOD!
ANOTHER
EXAMPLE

- o The results of our research have been published and made available to the industrial and scientific community.

EXPERIMENTS   7
ON A RXN. SYSTEM WHERE ESTABLISHED VALUES
NOT AVAILABLE

Time Line

We plan to carry out our research according to the project time line illustrated on the following page. The chart details every major step in the research project over a six month period.

Personnel

In order to perform the research, the only essential personnel required are Bob Miller, Sarah Rees and Betty Vazquez. However, in developing the computer interface with the reactor equipment, we may require an additional person with experience and knowledge of computer interface design.

Project Time Line

| Month 1 | Month 2 | Month 3 | Month 4 | Month 5 | Month 6 |

Good —
NICE PROFESSIONAL JOB

293

Requests

New Equipment

The following items are necessary for the continuation of the research project.

- o A desk top personal computer (286 machine)
- o Computer peripheral devices and software
- o An interface board
- o Constant temperature bath apparatus
- o Insulated reactor

Materials and Supplies

In conjunction with the above requested equipment, the following supplies are also requested.

- o A variety of chemicals and reagents
- o Computer diskettes
- o Miscellaneous glassware
- o Backup thermocouple

Support Services

As our research project is bench scale, the only support service needed on any regular basis is technical support for the computer interface.

Evaluation

Once we have established results from our research, we will submit our data for publication in the following journals:

- o Journal of the American Chemical Society
- o American Institute of Chemical Engineers Journal
- o Analytical Chemistry
- o Journal of Physical Chemistry
- o Review of Scientific Instruments

These publications will provide the necessary degree of coverage of our research. In addition to publishing our results, we intend to participate in the American Institute of Chemical Engineers convention in the winter of 1991.

Assessment Criteria for the Proposal Request for Funding

Below are listed the criteria for a *proposal request for funding*. After reading the exhibit, please circle the response indicating your judgment as to whether the group has exhibited

4 - Superior ability

3 - Competent ability

2 - Somewhat competent ability

1 - Lack of competent ability

Title Page

1. The project title, the investigators' names and addresses, and the institutional affiliation have been listed.

4   3   2   1

Table of Contents (one page)

2. All sections of the report have been listed, followed by the correct page designation.

4   3   2   1

3. The exact wording used for the table of contents has been used in each heading of the report.

4   3   2   1

Executive Summary (100 words)

4. The nature of the research has been described and the specific request has been defined.

4   3   2   1

Introduction

5. The research problem, its significance, and the expected gains have been described.

4   3   2   1

Background

6. The historical, theoretical, and industrial relevance of the research has been provided; the literature has been reviewed; and the present results have been reviewed.

4   3   2   1

Proposed Technical Solution

7. The proposed plan of research has been specified, the research goals have been identified, and the expected results have been proposed.

4   3   2   1

Management

8. A plan has been formulated for completion of the research, and a time line has been established.

4   3   2   1

9. A list of personnel needed to perform the research has been submitted.

4   3   2   1

Requests

10. Requisite new equipment, materials and supplies, and support services have been identified.

4   3   2   1

Evaluation

11. The anticipated means of publication, presentation, review, and evaluation of your research has been described.

4   3   2   1

References

12. Literature citations are provided.

4   3   2   1

Discussion Questions for the Proposal Request for Funding

1. How is the Executive Summary in the exhibit different from the Abstract in the formal laboratory report?

2. We often speak about an ethos of an engineering group, the particular personality and expertise of the group. If you were a funding agency, would you honor the request for funding made by the authors of this Request? On what basis would you award or deny the funding.

3. Ask your instructor to bring into class a call for proposals from the National Science Foundation. How are the specifications for proposals similar to those used in the exhibit? Ask your instructor about the planning process that goes into an actual request for funding. What does your instructor believe are the criteria for a successful request for funding?

4. The authors showed good insight in discussing the significance of the research. Explain. Why is this critical to the whole premise for this request for funding of new research?

5. In the section on industrial relevance, the authors failed to offer any concrete examples of reactive systems. How does this hinder the effectiveness of their Proposal?

6. In the section on fulfillment of the research, the authors desire to determine kinetic parameters and compare them to literature values. This is fine if the goal is simply to improve the apparatus. Based on the research goals listed in this proposal, are the authors displaying a true research orientation as described in Chapter 3?

7. In the section on fulfillment of the research, we suggest that the authors experimentally determine kinetic parameters of reactions which have not been studied extensively, or at all. Why would this add an exciting dimension to the proposal?

[1] A. E. Moorehead, "Proposals: The Process and the Document," in Charles H. Sides, ed., *Technical and Business Communication: Bibliographic Essays for Teachers and Corporate Trainers*, Urbana and Washington: NCTE and STS, 1989, pp. 265-286.

The Scholarly Paper

Introduction

We turn now to the last written reporting structure. The scholarly paper might seem to be an inappropriate document for undergraduate students. We believe, however, that it may serve as an introduction to graduate study, an aim of the approach used in this textbook.

Moreover, the scholarly paper is the embodiment of engineering research. Hence, it is concise, intense, and demanding. Theoretical relationships must be linked closely to results, results linked to discussion, and discussion to conclusions. An intensified form of the formal laboratory report, the scholarly paper requires you to think critically about the true nature of your experiment as an expression of the chemical engineering paradigm described in Chapter 3.

Organization of the Scholarly Paper

Organization of the Scholarly Paper

Title Page

Provide the abstract of your paper here.

Introduction

State the problem under examination.

Provide a review of the literature.

Conclude this section with a generative statement that will allow your readers to follow the pattern of your paper as you unfold your research.

Theory

Provide the most relevant theoretical relationships that you will draw on in your research.

Illustrate clearly what you are going to measure in light of the applicable theories.

Methods

Describe the equipment you used and include an original diagram.

Delineate the procedure you followed to obtain your information.

Include an analysis of any limitations of the equipment.

Discussion of Results

Present only that data that bears on the aim of your research; that is, provide the important figures and tables. All supplementary calculations are to be placed in the appendix.

Interpret the data you have just provided in terms of the fundamental principles described in your theory section.

Provide reasonable explanations for significant departures in your results from the theoretical model.

(Tactical note: Many scholarly papers discuss the results as they are presented. Note the difference between this technique and that used in the Formal Laboratory Report.)

Conclusion

Look beyond the results of your experiment and illustrate that you understand the nature and implications of your work.

Directions for Further Research

Discuss that which has intrigued you in this experiment.

Point towards future research and its implications.

References

Cite your sources.

Appendix

Provide your sample calculations, supplementary plots, and recommendations for equipment alignment.

Description of the Experiment

In the experiment on which the exhibit is based, a Karr Column used to perform a liquid-liquid extraction is evaluated. The primary objective is to determine how the extraction capability of the column varies with the mechanical agitation rate.

The available three-phase chemical system with which to evaluate the column is acetic acid in n-hexanol and water. In the Karr column, the water will be dispersed into the organic phase. An acid-rich aqueous phase is withdrawn from the bottom, while an acid-lean organic phase is taken from the top.

7

EXPERIMENTAL METHODS

Separation of the binary mixture of methanol and water was carried out using a ten sieve tray continuous distillation column equipped with a total condenser and total reboiler. The feed was introduced to either tray three or five, depending on the operating conditions. A diagram showing the equipment configuration is presented in Figure 1.

Figure 1
Continuous Distillation Column

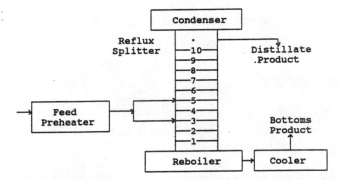

Good, Simple Schematic

The procedure used to obtain the experimental data involved the following:

With feed location of tray 5 chosen, cooling water at a rate of 17.6 gpm was flowed to the condenser and bottoms product cooler. A reflux ratio of 2:1 was set, and liquid feed was pumped into the column at a rate of 382.50 cm^3/min. Boilup was achieved using saturated steam at 5 psig and the distillate and bottoms product were recycled to the feed resevoir.

Flow rates not needed here

When steady state was established, and a reasonable temperature spread across the column realized, liquid and vapor samples were taken from each tray, condenser and reboiler. Samples were then analyzed with a refractometer operated isothermally at 25°C.

What is "reasonable"?

The entire procedure was repeated using tray 3 as the entry point under similar operating conditions.

There are certain operational limitations associated with the distillation column. Hydraulic limitations. such as flooding or weeping, prevent the usage of extremely high or low vapor flow rates, respectively. Consequently, inefficient mass transfer may result, and is commonly experienced when operating in the bubble and spray regions, for low and high vapor flow rates, respectively.

Generally, good! Not too much detail

← Inappropriate here

Sieve trays themselves are not very flexible, and usually must be operated at the designated operating conditions. Sieve trays are notorious for poor turndown, thus operating at significantly lower vapor rates than at the design specifications will result in lower efficiencies[20].

24

Condenser and Reboiler Duties

Theoretical reboiler and condenser heat loads were calculated and compared with experimental determinations. Theoretically, condenser cooling loads were calculated by[25]:

$$H' = (h_D + q_C/D)$$

and reboiler loads by[26]:

$$H' = (h_B - Q_R/B)$$

By neglecting heat loss to the surroundings, experimental condenser cooling loads and reboiler heat loads were calculated by making simple energy balances[27]:

$$Q = mc_p \Delta T$$

Results are summarized as follows:

	Feed Tray 5 (kJ/hr)	Feed Tray 3 (kJ/hr)
Condenser Duty		
theoretical	71,609	46,430
experimental	58,036	55,400
Reboiler Duty		
theoretical	-67,410	-47,198
experimental	-61,439	-58,649

For feed tray 5, the condenser duties were in error by 19%, and the reboiler duties by 8.9%. For feed tray 3, the condenser duties were in error by 16.2%, and the reboiler duties by 19.5%. It can be concluded that the correlations proved to be quite accurate in this experiment.

The differences observed between the theoretical and experimental condensor and reboiler duties may be attributed to the following:

- Experimental duties were based upon the assumption of adiabatic operation. Since heat loss to the surroundings was neglected a condenser or reboiler duty of greater magnitude resulted, therefore introducing a degree of error into the experimental results.

- Theoretical duties were determined from expressions derived from the Ponchon-Savarit method. Since the method requires graphical manipulations there are obvious errors in graphing points, determining scales, and drawing straight tie lines. These graphical difficulties thus result in erroneous condenser or reboiler duties.

GOOD DISCUSSION OF VARIATIONS. WHERE ARE UNCERTAINTY LIMITS?

301

25

CONCLUSION

What are your results really saying?

Although constant molal overflow was not strictly valid, overall column efficiencies calculated using the results of the McCabe-Thiele analysis compared favorably to those employing the results of the Ponchon-Savarit analysis. Location of the feed entry point had a marked effect upon overall column efficiency, and when moved from tray 5 to tray 3, column efficiency dropped from 40% to 30%.

The Murphree tray efficiency, E_M, varied substantially from tray to tray, ranging from 20% to 333%, and was considered to be an inaccurate measure of efficiency for this particular experiment.

DIRECTIONS FOR FURTHER RESEARCH

There are aspects of this experiment which deserve further research. First, instead of using a staged column for separating methanol and water, a packed column could have been used for distillation. Research in this area may show economic advantages and smaller pressure drops experienced when using a packed column over a staged column.

How do you know this?

Second, instead of using a reboiler, one can consider sparging saturated steam directly into the bottom of the column to provide boil-up. The use of direct or open steam raises questions of economic efficiency (no reboiler) as well as questions of improved separation. However, further research needs to be conducted in this area.

Third, since the distillation column control settings are optimally adjusted for feed entry at tray five, it would be desirable to obtain the optimal control settings for feed entry at trays three and seven. Operating the column under optimal settings for feed entry at these trays would provide a more accurate analysis of the separation efficiency for this particular distillation column.

Finally, one may conduct additional experiments using multiple feeds and/or a partial condenser.

These variations could possibly be useful in providing information on an industrial scale. Further research may lead to more inexpensive processes, more efficient separations, and perhaps a higher production capacity.

References
(a selection)

26

REFERENCES

1. Biddulph, Michael W.; Kalbassi, Mohammed A.
    "Distillation Efficiencies for Methanol/1-
    Propanol/Water". Industrial & Engineering
    Chemistry Research. v27. November 1988.
    p 2127-35.

2. McCabe W.L. and Thiele E.W. Industrial Engineering
    Chemistry. 1925. 17:605.

3. Wankat. Equilibriumn Staged Separations. Elsevier
    Publishing Co. New York. 1988. p 78.

4. McCabe Smith and Harriot. Unit Operations of Chemical
    Engineering. 3rd ed. McGraw-Hill. New York.
    1976. p 482

5. Wankat. Equilibriumn Staged Separations. Elsevier
    Publishing Co. New York. 1988. p 142

6. Ibid. p 130.

7. Geankopolis, Christie. Transport Processes and Unit
    Operations. 2nd ed. Allyn and Bacon Inc.
    Massachusets. 1983. p 641.

8. Ibid. p 641.

9. McCabe Smith and Harriot. Unit Operations of Chemical
    Engineering. 3rd ed. McGraw-Hill. New York.
    1976. p 487.

10. Felder and Rousseu. Elementary Principles in Chemical
    Processes. 2nd ed. John Wiley & Sons Inc. New
    York. 1986. p 104.

11. Geankopolis, Christie. Transport Processes and Unit
    Operations. 2nd ed. Allyn and Bacon Inc.
    Massachusets. 1983. p 660.

12. Ibid. p 661.

13. Ibid. p 662.

14. Ibid. p 662.

15. Ibid. p 663.

16. Wankat. Equilibriumn Staged Separations. Elsevier
    Publishing Co. New York. 1988. p 180.

Assessment Criteria for the Scholarly Paper

Below are listed the criteria for a *scholarly paper*. After reading the exhibit, please circle the response indicating your judgment as to whether the group has exhibited

4 - Superior ability

3 - Competent ability

2 - Somewhat competent ability

1 - Lack of competent ability

Title Page

1. The abstract has been provided on the title page.

4   3   2   1

Introduction

2. A background statement has described the problem under examination.

4   3   2   1

3. A review of the literature has been provided in this section.

4   3   2   1

4. A generative statement has allowed readers to anticipate the pattern of the paper.

4   3   2   1

Theory

5. The most relevant theoretical relationships have been established and applied to the problem at hand.

4   3   2   1

Methods

6. The equipment has been described and an original diagram has been included.

4   3   2   1

7. The experimental procedure has been described.

4   3   2   1

8. Any experimental limitations have been described.

4   3   2   1

9. Only that data that bears on the aim of the research has been presented.

4   3   2   1

10. The data are comprehensively interpreted.

4   3   2   1

Conclusion

11. The nature and broad implications of the work have been presented.

4   3   2   1

Directions for Further Research

12. Future research on an intriguing aspect of the experiment has been suggested.

4   3   2   1

References

13. Literature sources have been cited.

4   3   2   1

Appendix

14. Sample calculations, supplementary plots, and recommendations for equipment alignment have been presented.

4   3   2   1

Discussion Questions for the Scholarly Paper

1. In the Experimental Methods section of the Exhibit, why is the quotation of the flow rates of cooling water and distillate inappropriate?

2. Contrast the level of detail here with the Procedure sections of the User's Manual and Laboratory Report, especially in view of the targeted audiences for each. Why does the Scholarly Paper warrant the least amount of detail?

3. Why are the last two paragraphs in the Experimental Methods section not appropriate here? Where do they belong, if at all?

4. How might you rewrite the Conclusion?

5. The recommendations for Further Research are interesting, but they do not appear well developed. When offering such ideas, why would it be useful to be more quantitative, or to cite a reference from the literature?

The Oral Presentation Proposal

Introduction

Successful technical presentations require a strong technical content. When working on a presentation, always make sure that the technical content is of the highest quality possible. Make sure you really understand the material you are presenting. It is usually easy for an audience to spot a weak technical presenter. A thorough understanding of the material you are presenting is the best insurance against embarrassment.

The structure of the oral presentation often follows the same reporting structure outlined in the formal laboratory report. We recommend that you follow the same basic guidelines. If practical, distributing a handout to the audience just prior to the oral presentation is recommended. This handout is usually made up of hard copies of the slides to be used (in order), as well as supporting material in an Appendix. Audience members often find such handouts especially useful for jotting down notes as you speak.

However, an established format will not, by itself, guarantee an effective oral presentation. Even technically robust presentations can lose their effectiveness if the presenter does not offer the material clearly, logically, and, perhaps most importantly, on a level appropriate to the audience's expertise. Always make sure you know your audience and the technical background of the people you are addressing. (For audience analysis strategy, see Chapter 9.)

Furthermore, the quality of the presentation can be greatly enhanced by other, more subtle factors—using visual aids that are easy to understand and visually satisfying, speaking with good tone and modulation, assuming an air of professionalism, employing effective rhetorical devices (e.g., good eye contact, reinforcing gestures, relevant examples, insightful analogies.) Some of these elements are covered in Chapter 9's Sidebar on tips for an effective oral presentation.

Good presentation skills can be mastered by a willingness to learn, practice, and experience. The payback is typically very significant.

Organization of the Oral Presentation

The organization of the oral presentation in the laboratory typically follows the organizational structure of the formal laboratory report. (Refer back to the discussion of that reporting structure in this chapter.) The overview transparency in the Exhibit indicates that structure.

In other settings, the organization of the oral presentation varies according to the audience's informational needs.

Description of the Experiment

In this experiment, reactor engineering is applied to the biological reactions of yeast cell growth. The objective is to determine the kinetics of yeast growth. The experimental objectives include the determination of cell mass and glucose concentrations as functions of time.

OVERVIEW

BACKGROUND

INTRODUCTION

APPARATUS

PROCEDURE

THEORY

RESULTS

CONCLUSIONS

RECOMMENDATIONS

GOOD - AUDIENCE MUST BE TOLD WHAT IS COMING UP

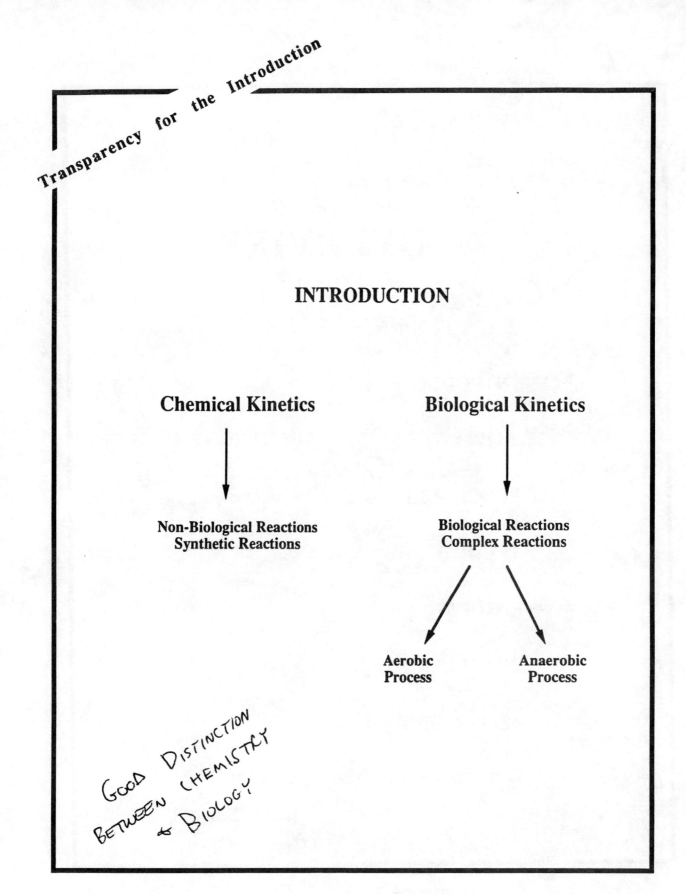

Theory

Yeast Growth Characteristics

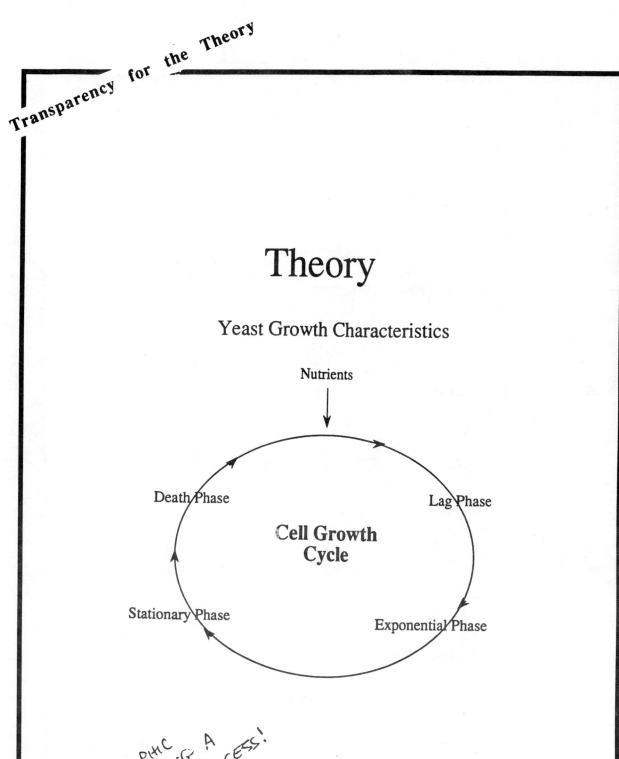

Nutrients

Death Phase

Lag Phase

Cell Growth Cycle

Stationary Phase

Exponential Phase

GRAPHIC DESCRIBING A COMPLEX PROCESS!

Theory (cont'd)

● Generation Time   (T)

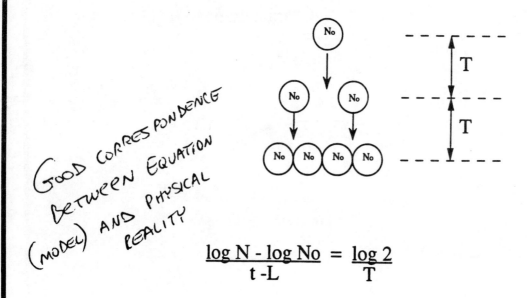

GOOD CORRESPONDENCE BETWEEN EQUATION (MODEL) AND PHYSICAL REALITY

$$\frac{\log N - \log No}{t - L} = \frac{\log 2}{T}$$

T = time / generations

t = time to reach final cell concentration

L = lag time

Theory (cont'd)

Analytical Method

- Number of Generations

$$N_o \longrightarrow 2N_o$$

$$N_o \longrightarrow 4N_o$$

$$N_o \longrightarrow 2^z N_o$$

$$N = 2^z N_o$$

$$Z = \frac{\log N - \log N_o}{\log 2}$$

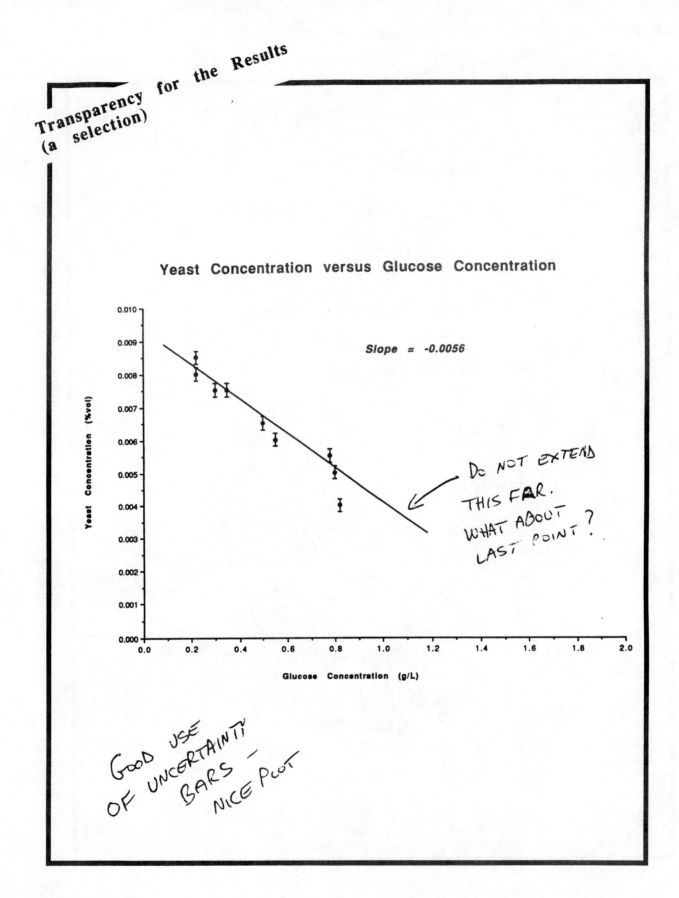

Assessment Criteria for the Oral Presentation Proposal

Below are listed the criteria for an *oral presentation proposal*. After observing the group's presentation of a given experiment, please circle the response indicating your judgment as to whether the group has exhibited

4 - Superior ability

3 - Competent ability

2 - Somewhat competent ability

1 - Lack of competent ability

Evidence of Planning the Presentation

1. The audience has been identified and analyzed, and the presentation has been prepared in terms of its needs.

   4   3   2   1

2. The presentational strengths of the members of the group have been well employed.

   4   3   2   1

3. An overall presentation plan has been developed that demonstrates the pattern of the research.

   4   3   2   1

4. An effective presentation proposal packet (handout) has been prepared.

   4   3   2   1

5. Overheads have been prepared that lead the audience through the presentation.

   4   3   2   1

Evidence of Logic and Critical Thinking

6. Presenters organized their ideas into logical sequences.

   4   3   2   1

7. Presenters offered their ideas along with the ideas of others through the citation of sources.

   4   3   2   1

8. Presenters analyzed and understood the limits of their experimental data.

   4   3   2   1

9. Presenters illustrated an understanding of the laws, theories, and hypotheses relevant to their research through the collection and analysis of data, and they determined if their data support or contradict these laws, theories, and hypotheses.

   4   3   2   1

313

10. Presenters displayed data in a way that reveals relationships between variables.

4   3   2   1

11. Presenters were able to formulate mathematical problems in their field and to apply fundamental analytic, numerical, and computational techniques for solving them.

4   3   2   1

12. Presenters were familiar with the manner in which a technological problem is formulated and solved.

4   3   2   1

13. Overall Score:

4   3   2   1

Discussion Questions for the Oral Presentation Proposal

1. Why is a simple graphic, such as the one shown in the Theory section of the Exhibit, so useful in describing a complex process, especially for a mixed audience in an oral presentation?

2. Discuss the usefulness of using a graphic to describe an analytic model, such as was used in the Theory section. Why is this so powerful for achieving effective communication (translation) with a mixed audience during an oral presentation?

3. Why do the uncertainty bars make the plot of yeast concentration vs. glucose concentration so effective? Notice that the authors have not rejected the last, outlying date point, but they have extended their regressed line too far. What else could they have done?

An Essay for Instructors

Education . . . may actually train automatic thinking. The reason is that so much of what passes for education consists of memorizing connections between words, phrases, and images. These words, phrases, and images may or may not, in turn, have any link to reality.

Robin M. Dawes, *Rational Choice in an Uncertain World*

Basically, "Grandfather's Lab" has always been the educational medium in which the student has been given the opportunity to put into practice the theoretical skills learned in the classroom. In this process, the student is expected to learn the "real world" of engineering by being subjected to a number of significant exercises to develop his [or her] skills as a practicing engineer. The basic philosophy behind these laboratories was very good for times past and one which was strongly endorsed by the majority of engineering educators. In today's economy, however, this just won't "cut it." There are a number of problems associated with the Grandfather's Lab that need to be addressed in the near future if the laboratory experience is to remain a viable part of the undergraduate engineering curriculum. We must ask ourselves: "How can our laboratories, in their present state, possibly prepare our students for a career in a time of such unprecedented dynamic change as we have today?"

The answer is: "They can't."

Sam Hilborn, "Can We Afford our Grandfathers' Laboratories?"

The Foundations of the Textbook

Recent papers and reports reveal that four broad goals must be met if we are to empower our students to remain competitive.[1] Each of these goals has important implications for higher education.

First, the United States must be prepared for integration into a global economy bound to international trade. Thus, our students must demonstrate a respect for cultural diversity while retaining the democratic values of our nation. Second, scientific discovery and technological development will continue to increase at a rapid rate. Thus, our students will need to have a research-orientation to their careers that will result in life-long learning. Third, new technologies and the environments in which they will exist will be developed along interdisciplinary boundaries. Thus, those of us who plan and implement the curriculum must recognize that the answers may be ahead of, not behind, us. Fourth, the role of communications and critical thinking will be key if our students are to be vehicles of change and of innovation.

What, then, is needed in today's chemical engineering undergraduate curriculum? Based on our review of the literature, our industrial experience, and our academic experience, we offer four principles below which are the foundations of this textbook.

Principle 1. Interdisciplinarity is essential to successful chemical engineering practice. As Ralph Landau argues in his historically-based analysis,[2]

The need for an integrated chemical engineering education—with a total systems approach and an emphasis on design—is greater than ever. This is for both old and new fields of application. Thus, not only is the scale-up skill important, so is the ability to lead a diverse team of specialists while retaining an overall economic marketing and technical perspective. If we cannot convey this to prospective new chemical engineers, we risk the loss of our identification and the employment contest . . . . We must become more aware of the larger environment that shapes our practices and industries.

Hence, a new level of expertise is demanded. In this textbook, our special contribution to the call for greater interdisciplinarity is to introduce chemical engineering students to recent advances in the fields of critical thinking and technical communications.

Principle 2. The ability to think critically is essential to a successful career in chemical engineering. In the closing decade of the twentieth century, critical thinking must assume a central place in any field, but this is especially important in chemical engineering. As the National Research Council explains in its report *Frontiers of Chemical Engineering*,[3]

It is difficult to visualize the world without the large volume of production of antibiotics, fertilizers, agricultural chemicals, special polymers for biomedical devices, high strength polymer composites, and synthetic fibers and fabrics. . . . Chemical engineering, however, is more than a group of products or a pile of economic statistics. As an intellectual discipline, it has its characteristic set of problems and systematic solutions for solving them; that is, its paradigm. Since the birth of chemical engineering in the last century, its fundamental paradigm has gone through a series of dramatic changes, and more are on the way.

The ability to think creatively in disciplinary contexts is presently a matter of national significance. If, as instructors, we are to produce citizens capable of meeting the demanding tasks found in the modern marketplace, then we need to promote a kind of professionalism which allows students to excel, not merely survive.

Principle 3. The ability to communicate effectively is essential to a successful career in chemical engineering. The need of strong communication skills is apparent in a well-known survey of 4,759 readers of *Chemical Engineering*. The survey reveals that although U.S. educated chemical engineers are fairly well prepared for their first jobs after graduation, many lack the communication skills needed for success on the job. "Graduates," the survey found, "are unaware of, and underprepared for, the vast amount of communicating (writing and speaking) they will have to do in industry." When asked what subjects they found useful in their careers, the top subject—ahead of unit operations, process design, transport phenomena, process chemistry, and transport phenomenon—was communications. 87.8% of those surveyed answered that training in written communication was useful, while 87.2% answered that training in oral communication was useful. The study of material and energy balances placed a poor fifth at 78.9%.[4]

Because communication is important to success, students must have a great deal of practice with the aims and forms of communication they will encounter in organizational settings.

Principle 4. The ability to understand interrelationships in science, technology, and society is essential to a successful career in chemical engineering. The study of science, technology, and society allows students to analyze the relationship among what Arnold Pacey has termed in *The Culture of Technology* the technical, organizational, and cultural aspects of technology practice.[5]

The application of these principles allow what we have termed a critical thinking approach. Stated briefly, critical thinking can be defined as the ability to achieve independent thought, intellectual breadth, cultural breadth, and ethical awareness. The most powerful vehicle for the achievement of critical thinking is writing ability, a tool which allows students to record, examine, and communicate their ideas. It is the centerpiece of our model.

The Development of the Textbook

We began our project by familiarizing ourselves with scholarship that was accessible to us as an interdisciplinary team, especially that scholarship which has recently emerged on critical thinking.[6]

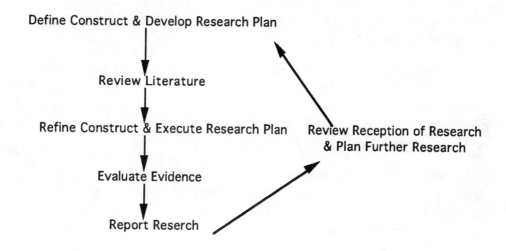

Figure 1. The Research Plan that Produced *The Compleat Chemical Engineer*

This provided us with the foundations upon which to construct a research plan, which is illustrated in Figure 1 below. This textbook is the culmination of the plan.

Our research plan, although complex in its execution, was straightforward in its aim. We wanted to identify those behavioral patterns that allow students to be successful chemical engineers. The two-semester senior chemical engineering laboratory course provided the perfect "laboratory" for our research since in these courses the students are expected to rely upon all their previous training in a capstone effort. Where desirable behavior existed, it would be reinforced. Where it was found lacking, corrective measures would be taken with the individual students. In addition, if the deficiency proved to be institutional within the class, then recommendations would be made for corrective action in the prior curriculum.

In the laboratory we observed our students closely by watching them perform during actual experimentation sessions, write draft reports and revise them, and make oral presentations. While we were clearly not trained ethnographers, we did employ many of the methods of close observation.[7]

We also examined the contents of chemical engineering laboratory courses in twenty universities.[8] Three patterns emerged. First, although there was variation, each department had core experiments in unit operations and chemical reactions. Second, instructors' handouts were used to describe the experiments. Third, there was no formal textbook used to strengthen the abilities of students to think critically and communicate effectively. We concluded that, as we had assumed, there was a common content to the laboratory course. We also concluded that there was a need for a thorough treatment of communications and critical thinking within this specialized laboratory setting. This is one of the primary motivations for this textbook.

During our research, it became apparent that our graduating seniors would have an even higher likelihood of success as critical thinkers if critical thinking and communications became an integral part of the undergraduate curriculum. To this end, we developed new sophomore-level courses to give new chemical engineering students an introduction to their chosen profession on a broad level. These "Professional Development" courses are yoked to the first material and energy balance calculation courses. Using experiential and disciplinary focuses, these courses introduce critical thinking and communications skills.

Students who have taken the above courses as sophomores and who have been exposed to other elements of critical thinking being introduced into other courses have performed at superior levels in the senior laboratory course. Hence, we decided to broaden this textbook beyond the laboratory course to include the sophomore courses, including plant design. Our textbook, it thus appears, could find a home within chemical engineering.

In summary, we conducted a number of qualitative and quantitative studies in order to write this book. Because of our research-based approach, you should feel confident that the topics we cover and the suggestions that we make are based on the same kinds of rigor that characterize chemical engineering research itself.[9]

Special Features of the Textbook

The research method described above has allowed us to incorporate six unique features into the manuscript. We believe these features make the textbook quite useful for students.

Examples from Chemical Engineering. Because we have studied the critical thinking processes of students within various chemical engineering courses, all of our examples are extraordinarily specific in addressing the kinds of problems and solutions students will actually encounter.

Visual Models. One aspect of our research involved the design of models of the ways that we wanted our students to think. As a result, figures accompany all discussions of critical thinking. Students can also tailor the models to their own specifications, and instructors can discuss their potential strengths.

Reporting Structures and Exhibits. Research in the field of communications tells us that both students and graduates must be able to communicate their findings to a variety of audiences through a variety of means. Therefore, there are guides to numerous ways for students to record and report information. In addition, we provide samples, or exhibits, of the ways that students approach these various forms of writing. Taken from the work of our students, these exhibits are reproduced and annotated with our comments.

Assessment Criteria. We have specified criteria for each reporting structure and devised a four-point scoring system for these criteria. Both students and their instructors will be able to use these scales to identify problem areas and to measure improvement. These scales can be used to verify student performance for program accreditation purposes.

Brief, Interesting Exercises. To enhance the students' experiences in their chemical engineering courses, we have provided a variety of brief, engaging exercises in each chapter that will allow the instructor to discuss, for example, the need for applied ethics or the ways to pursue engineering argumentation.

A Scholarly Approach. A series of extensive notes unifies the text. These notes, written with the student in mind, will stimulate thought and allow further study. As well, these notes will remind students that the textbook is based on research and is not merely a random scattering of clever ideas.

The Audience for this Textbook

We assume that this textbook will be used primarily by all students majoring in chemical engineering. In fact, we see this book as a chemical engineering student's "Vade Mecum".[10] Due to its emphasis on critical thinking and communications, this text could also be used by any science or engineering student, as well as by professionals in the work place.

This textbook is thus intended to be, in effect, a handbook for students on critical thinking in chemical engineering which can be used throughout their professional education. Instructors can use it for specific courses as the primary or as a secondary textbook.

A Syllabus for Using this Textbook with Your Students in Introductory Courses

If your are an instructor of beginning chemical engineering students, we offer the following syllabi for a two-semester sequence of one-credit courses intended for the sophomore year. (In our university, these courses are joined to a two-semester sequence in material end energy balances.) In addition to serving as an introduction to the profession of chemical engineering, the sequence serves to begin formal training in critical thinking, including communications, in chemical engineering.

Figure 2. Sample Chemical Engineering Professional Development Course Plans Incorporating *The Compleat Chemical Engineer*

Chemical Engineering 224
Topics and Schedule for the Fall

Instructor: N. Elliot, Department of Social Science and Policy Studies

Course Description:

| ChE 224 | Focus: Experiential | Description: Students explore their own perceptions in relation to the field of chemical engineering | Aim: To enable students to investigate their academic interests and professional goals |

In this one-credit course students are invited to examine the circumstances that brought them to the field of chemical engineering. As well, students are asked to plan their progress through the discipline.

Ethics plays a major part in the course. Discussion focuses on the framework for the analysis of ethical theories as that framework arises in two case studies: penicillin production and the explosion of the Challenger.

The major assignment centers around an experientially-based paper in which students write about the paths that led to their major, their areas of interest in the field, their potential career expectations, and their academic and professional goals.

Syllabus:

Meeting 1

Topics:

An STS Program Development Model for Chemical Engineering
Critical Thinking in Chemical Engineering
The Experiential Response: Primo Levi's *The Periodic Table*

Chapter 1 Assignment for Sophomores: The Facts (*The Compleat Chemical Engineer*)

Meeting 2

Topics:

Understanding the Tools for Ethical Decision Making
The Nuts and Bolts of Technology
Applied Ethics I: Once a Great Moment—Penicillin Production
Applied Ethics II: Once A Tragic Moment—Roger Boisjoly and the Challenger

Draft of ChE 224 Paper #1—Historical Analysis of Penicillin Production and the Challenger Explosion. Read Chapters 2 and 10 of *The Compleat Chemical Engineer*. The paper should follow the form of the historical research paper described in Chapter 10. Use APA format.

Meeting 3

In-Class peer review of ChE 224 Paper #1

Begin to work on ChE 224 Paper #2—Experiential Paper. Paper #2 should follow the form of the experiential paper described in Chapter 10 of *The Compleat Chemical Engineer*. Use MLA format.

Meeting 4

ChE 224 Paper #1 Due: Historical Analysis of Penicillin Production and the Challenger Explosion

Prepare for the Experiential Paper

Meeting 5

ChE 224 Paper #2 Due: Experiential Paper

Chemical Engineering 226
Topics and Schedule for the Spring

Instructor: N. Elliot, Department of Social Science and Policy Studies

Course Description:

ChE 226	Focus: Disciplinary	Description: Students explore the role of the chemical engineer in the work place	Aim: To enable students to investigate the organizational role of the engineer

In this second one-credit sophomore level course, students explore the range of careers in their field. To this end, discussion focuses on the history and development of the profession: the beginning of unit operations by A.D. Little, the connection between chemical theory and engineering practice early in the century, the emergence of chemical engineering as a model profession during World War II, and the current emphasis on chemical engineering as a globalized, interdisciplinary profession. Applied value studies are encouraged as guest lecturers discuss government regulations.

The major assignment requires students to perform a case study of a chemical engineer. The choice of subject, usually an alumni of our program, helps develop program cohesiveness. Students are asked to develop an interview methodology, utilize that methodology in the interview, analyze the respondent's comments, infer a portrait of a working professional, and design directions for further research. Because the overall design of this assignment follows the core engineering investigative methodology itself, students begin to see potential relationships between the social sciences and engineering.

Syllabus:

Meeting 1

Topics:

> Notes on American Technological History from 1870–1920
> Identify Concept of interest for Field Interview

Meeting 2

Topics:

> Background to Case Study of a Chemical Engineer
> Technologizing America

Prepare the historical research paper according to specifications in Chapter 10 of *The Compleat Chemical Engineer*. Use ACS format.

Read Chapter 10 on the field research paper. Concept of interest and subject identified for Case Study by Meeting 3. Use either APA or ACS format.

Meeting 3

ChE 226 Paper #1 Due: Issues in the History of Chemical Engineering

Peer and Instructor Review for Field Study

Meeting 4

ChE 226 Paper #2 Due: Case Study of a Chemical Engineer

A Syllabus for Using this Textbook with Your Students in Advanced Courses

If you teach the capstone laboratory course to juniors and seniors, you probably distribute to your class a series of technical handouts that explain in detail the experiments in your laboratory. *Our textbook is intended as a supplement to these handouts.* We assume that you are interested in the kinds of topics we address in this book.

The example syllabus below should help you integrate this book into your existing course.

Figure 3. Sample Chemical Engineering Laboratory Course Plan Incorporating
The Compleat Chemical Engineer

Course Title: ChE 487 - Chemical Engineering Laboratory I

Credits: 4

Meeting Time: Thursdays, 10 am —> 4 pm

Prerequisites: ChE 363 (Transport Operations I), ChE 364 (Transport Operations II), and Math 225 (Survey of Probability and Statistics)

Instructor: Robert Barat

Texts: *ChE 487 Laboratory Manual, 1992 Version* (photocopied guides to the experiments)

R. Barat and N. Elliot, *The Compleat Chemical Engineer*

Organization: During the first meeting, students will organize themselves into groups of three persons each.

Procedure for Each Laboratory Session: During the first twenty minutes of the laboratory session, I will discuss a topic from the Barat and Elliot textbook: communications, error analysis, the ethical use of data, and so forth. I will then review general questions that the class will have about the experiments for another 10 minutes. After this initial thirty-minute session, each group will then spend the remainder of the session working on the experiment at hand. Before beginning any new experiment, however, I will hold a planning conference with each group.

Purpose of Course: This is the capstone course in chemical engineering. In light of that fact, it will provide you with a number of opportunities to pull together your undergraduate experiences in many of the areas you have studied both within and beyond the field of chemical engineering. It will also provide you with a chance to strengthen your problem solving and communication strategies in chemical engineering.

Grades: The final grade in this course will be based on the progression of your work throughout the semester. Grades will be determined in three areas:

—10% Brief exercises (taken from the end of the chapter in Barat and Elliot and/or my own exercises; more information on these exercises will be provided as the course proceeds)

—60% Labs presented in various reporting structures

—20% Final Interdisciplinary Oral Presentation

—10% Laboratory Notebook

A Note on Exercises: Most exercises will be taken from the supplemental book you are to purchase from the bookstore: R. Barat and N. Elliot's *The Compleat Chemical Engineer*. Additional oral and written exercises will also be provided.

A Note on Drafts: As you will see on the syllabus below, each group must submit a polished draft of each paper (i.e., word processed, all tables and figures presented, all data available). I will then review this draft with the group and suggest revisions that will turn the report into a stronger final product. A preliminary grade will be placed on the draft at that time. When the draft is revised, that grade may be raised only if substantial improvement has been made. Any group not turning in competent work on a draft (i.e., at least a grade of C) will not be allowed to revise the draft, so do your best work on all drafts.

A Note on the Interdisciplinary Oral Presentation: During the last class meeting, you will present your last laboratory to an interdisciplinary audience with specialities in chemical engineering, chemistry, environmental engineering, environmental history, and technical communication. This thirty-minute report will be scored according to criteria that you will be given in advance.

Week-by-Week Syllabus

Week 1

Introduction to Course
Assignment of Laboratory Groups
Assignment: Skim Chapter 1 (Thinking Critically in Chemical Engineering), Chapter 5 (Conducting the Literature Search) and Chapters 9–11 (On Communication). Read carefully the discussion of safety in Chapter 3.
Complete the Chapter 1 Assignment, Surveying Your Progress
Complete the Chapter 7 Assignment, Analyzing Laboratories

Week 2

Discussion of the following experiments for the course: fluid flow, pressure drop and liquid holdup in packed towers, continuous heat transfer, and batch heat transfer. Discussion of the following four Reporting Structures for the course (From Chapter 11):
    —The Formal Laboratory Report
    —The Procedural Manual
    —Request for Funding
    —Oral Presentation Proposal

Discussion of assessment in the course
Begin Laboratory 1
Read Chapter 2 (Interpreting the History of Chemical Engineering)

Week 3

Discussion of the History of Chemical Engineering
Read Chapter 3 (Working in the Laboratory)
Complete Assignments 1, 2, 4, and 6

Week 4

Discussion of A Research Plan for Chemical Engineering
Complete Laboratory 1
Read Chapter 4 (The Uses of Argument) and Chapter 5 (Conducting the Literature Search)
Complete Assignments 1, 2, and 3 for Juniors and Seniors for Understanding the Literature Search

Week 5

Discussion of Argumentation in Engineering
Begin Laboratory 2
Draft of Laboratory 1 (written as a Formal Laboratory Report) due; see Chapter 11 for specifications of this reporting structure
Read Chapter 9 (Communicating Information in Chemical Engineering)
Complete Chapter 4 assignment, Aanalyzing an Argument in Chemical Engineering

Week 6

Discussion of Communicating Findings in Chemical Engineering
Final Copy of Laboratory 1 (written as a Formal Laboratory Report) due
Read Chapter 6 (Ethical Decision Making in Chemical Engineering)
Complete Assignment, Laboratory Ethics and Outlying Data Points

Week 7

Discussion of Ethical Decision Making in Chemical Engineering
Complete Laboratory 2
Read Chapter 7 (Planning the Laboratory Environment)
Complete Assignments 1, 2, and 3—Analyzing Laboratories

Week 8

Discussion of Planning the Laboratory Environment
Begin Laboratory 3
Draft of Laboratory 2 (written as a Procedural Manual) due;
see Chapter 11 for specifications on this reporting structure

Week 9

Discussion of Appropriate Journals and Magazines for the Professional
Final Copy of Laboratory 2 (written as a Procedural Manual) due

Week 10

Complete Laboratory 3

Week 11

Begin Laboratory 4
Draft of Laboratory 3 (written as a Request for Funding) due; see Chapter 11 for specifications on this reporting structure

Week 12

Discussion: Preparing for an Interdisciplinary Presentation
Final Copy of Laboratory 3 (written as a Request for Funding) due

Week 13

Discussion of Chemical Engineering in the 21st Century
Complete Laboratory 4

Week 14

Draft of Laboratory 4 (designed as an Oral Presentation Proposal) due
Dry Runs for Final Oral Presentation (30 minutes per group); see Chapter 11 for specifications on this oral reporting structure (the presentation proposal)

Week 15

Final Oral Presentations (all groups)

As you can see, this syllabus allows you to broaden the scope of the course while retaining its fundamental nature. Of course, other strategies can be used depending on the interests of the class and the point of view of the instructor. Regardless of the strategies or emphasis, however, by using this book, you and your students will experiment with critical thinking as you experiment with chemical engineering principles in the laboratory.

A Closing Statement

An objection may be sounded for using the approach we advocate in this book. It may be argued that such an emphasis will weaken the technical content of the chemical engineering courses. We have not observed this. Indeed, in writing and speaking so frequently about their experiments and other assignments, students come to think reflectively about them. Instead of being a burden, the great emphasis upon communications actually enables students to strengthen their thinking abilities. We have discovered, in fact, that our approach helps students articulate many assumptions about the field that they have made but never tested, such as research methodology or the best tactics to present an argument. Our approach helps students synthesize their education.

In essence, this book is about taking charge. It is not about elaborating the instructional difficulties or exaggerating academic problems to such a degree that change is impossible. As we prepare our students for a world that they will shape—a world that was, in many ways, unimaginable just twenty years ago—we must create an educational environment that will ensure both their technical competency and their thinking abilities. This textbook is a step in that direction.[11]

Notes and References

1. Sam Hilborn, "Can We Afford our Grandfather's Laboratories?" *Proceedings: Innovation in Undergraduate Engineering Education II, Engineering Foundation Conferences,* 1990, pp. 1–6; Juris Vagners, "Undergraduate Control Systems Engineering Design Revisited," *Proceedings: Innovation in Undergraduate Engineering Education II, Engineering Foundation Conferences,* 1990, pp. 23–37; Joseph S. Johnston, Jr., Susan Shamen, and Robert Zemsky, *Unfinished Design: The Humanities and Social Sciences in Undergraduate Engineering Education,* Washington, D.C.: Association of American Colleges, 1988; Committee on the Education and Utilization of the Engineer, Commission in Engineering and Technical Systems, National Research Council, *Engineering Education and Practice in the United States: Foundations of Our Techno-Economic Future,* Washington, D.C.: National Academy Press, 1985; American Association for the Advancement of Science, *Science for All Americans: Summary of Project 2061,* Washington, D.C.: American Association for the Advancement of Science, 1989.

2. Ralph Landau, "The Chemical Engineer and the CPI: Reading the Future from the Past," *Chemical Engineering Progress,* September 1989, pp. 25–39.

3. National Research Council, *Frontiers of Chemical Engineering,* Washington: The National Academy Press, 1988.

4. Mark A. Lipowicz and Roy V. Hughson, "A *CE* Survey: Putting College Back on Course," *Chemical Engineering,* September 19, 1983, pp. 48–60.

5. Arnold Pacey, *The Culture of Technology*, Cambridge: MIT Press, 1983.

6. For helpful reviews of this work, see especially the following three studies: D. N. Perkins, *Knowledge as Design*, Hillsdale, New Jersey: Lawrence Earlbaum, 1986; Richard Paul, *Critical Thinking*, Rohnert Park, California: Center for Critical Thinking and Moral Critique, 1990; Lauren B. Resnick, *Education and Learning to Think*, Washington, D. C.: National Academy Press, 1987.

7. For a discussion of ethnographic methods of the kind we employed, see the following: Stephen Doheny-Farina and Lee Odell, "Ethnographic Research on Writing: Assumptions and Methodology," in *Writing in Nonacademic Settings*, eds. Lee Odell and Dixie Goswami, New York: Guilford Press, 1985, pp. 503–535; Dixie Goswami and Peter R. Stillman, eds., *Reclaiming the Classroom: Teacher Research as an Agency for Change*, Upper Montclair, New Jersey: Boynton/Cook, 1987.

8. California State Polytechnic University, Clarkson University, Colorado School of Mines, Hampton University, Manhattan College, Montana State University, Pennsylvania State University, Purdue University, State University of New York at Buffalo, Stevens Institute of Technology, TriState University, University of Alabama, University of Arizona, University of Arkansas, University of California at Berkeley, University of Dayton, University of Massachusetts at Amherst, University of Utah, University of Wisconsin, and the University of Washington. The inquiry was made by our colleague, Reginald Tomkins.

9. We have reported our on-going research as noted in the Preface.

10. "Vade Macum" (Latin, "go with me"); in effect, a useful handbook that a person always takes along.

11. Ours, however, is clearly not the first step but rather part of an AIChE tradition of curriculum involvement. In fact, this textbook may be seen as a specialized extension of the 1987 AIChE publication, *Chemical Engineering: What Skills are Needed?* For recent studies that call for a new look at chemical engineering education, see especially the following: Davor P. Sutija and John M. Prausnitz, "Chemical Engineering in the Spectrum of Knowledge," *Chemical Engineering Education*, Winter 1990, pp. 20–25; Mark Rosenzweig, "Are ChE Students as Good as We Were?" *Chemical Engineering Progress*, February 1991, p. 7; R. Russell Rinehart, "Improving the Quality of Chemical Engineering Education," *Chemical Engineering Progress*, August, 1991, pp. 67–70. For a broader assessment of new directions for scientific and engineering education, see James Krieger's "Winds of Revolution Sweep through Science Education," *Chemical and Engineering News*, June 11, 1991, pp. 27–43.